Bird Nests and Construction Behaviour provides a broad view of our current understanding of the biology of the nests, bowers and tools made by birds. It illustrates how, among vertebrates, the building abilities of birds are more impressive and consistent than those of any other builders, apart from ourselves, yet birds seem to require no special equipment, and use quite uncomplicated behaviour. In doing so, the book raises general issues in the field of behavioural ecology, including the costs of reproduction, sexual selection and the organisation and complexity of behaviour. Written for students and researchers of animal behaviour, behavioural ecology and ornithology, it will nevertheless make fascinating reading for architects and engineers interested in understanding how structures are created by animals.

MIKE HANSELL is Senior Lecturer in the Institute of Biomedical and Life Sciences, at the University of Glasgow. In his research career he has been interested in structures made by many different species, including caddis larval cases, wasp nests and mammal burrows, but his current interests concentrate on bird nests, bowers and tools. He is the author of *Animal Architecture and Building Behaviour* (1984).

Bird Nests
and Construction Behaviour

Mike Hansell

Pen and ink illustrations by
Raith Overhill

CAMBRIDGE
UNIVERSITY PRESS

CAMBRIDGE UNIVERSITY PRESS
Cambridge, New York, Melbourne, Madrid, Cape Town, Singapore, São Paulo

Cambridge University Press
The Edinburgh Building, Cambridge CB2 2RU, UK

Published in the United States of America by Cambridge University Press, New York

www.cambridge.org
Information on this title: www.cambridge.org/9780521460385

First published 2000
Reprinted 2002
This digitally printed first paperback version 2005

A catalogue record for this publication is available from the British Library

Library of Congress Cataloguing in Publication data

Hansell, Michael H. (Michael Henry), 1940–
    Bird nests and construction behaviour / Mike Hansell; pen and ink
illustrations, Raith Overhill.
        p. cm.
    ISBN 0 521 46038 7 (hb)
    1. Birds – Nests – Design and construction. 2. Birds – Behaviour. I. Title.
QL675.H34 2000
598.156′4 – dc21   99-087681

ISBN-13 978-0-521-46038-5 hardback
ISBN-10 0-521-46038-7 hardback

ISBN-13 978-0-521-01764-0 paperback
ISBN-10 0-521-01764-5 paperback

To *Norma, Christopher* and *Lindsay*

Contents

Acknowledgements

My research interests embrace all animal architecture and building behaviour but, until recent years, I had confined my personal research to insect builders. Inevitably, as it now seems, I was drawn to the study of bird nests, because so much about them remains to be understood, the structures themselves have a tangible appeal, and the research environment at Glasgow University is particularly strong in various aspects of ornithology. Without this climate of support from colleagues I would not have had the confidence to undertake this book, or certainly not one with this breadth of coverage.

To give this book added authority I felt it was necessary to see and handle a large number of nests. The answer was to study museum collections. I am therefore most grateful to the following museums for their co-operation: Kelvingrove Gallery, Glasgow; Royal Scottish Museum, Edinburgh; Musée d'Histoire Naturelle, Rouen. Longer visits were necessary to the major collections at Musée National d'Histoire Naturelle, Paris; the Natural History Museum, London; the National Museum of Natural History, Washington; and the Western Foundation of Vertebrate Zoology, California. I am particularly indebted to Lloyd Kiff at the Western Foundation for the personal attention he gave to my study of that fine collection. To enable me to visit these places, I am most grateful for the financial support given to me by the Association for the Study of Animal Behaviour and by the Carnegie Trust for Scottish Universities.

In writing the book I have been fortunate to be able to draw upon the considerable expertise of colleagues who were kind enough to find time to make constructive comments on chapters in draft and draw my attention to important literature of which I was unaware. In particular I would like to thank Ian Barber, David Houston, Felicity Huntingford, Neil Metcalfe, Rudi Náger, Rod Page and Graham Ruxton. The book is much improved by their valuable contributions. I am also indebted to Stuart Humphries and Graham Ruxton for statistical analysis of nest profile data. Many others have helped in smaller ways in conversations over coffee or wherever; my thanks to them also.

I have been sent quite a number of nests from Britain and around the world by a now quite extensive list of bird enthusiasts and researchers. This has enabled me to examine some nests at leisure and in a detail not possible in museum visits. It is a real pleasure to receive this material and, although much remains to be done with it, it has already made its contribution to this book.

I would like to express my gratitude to Raith Overhill for his skill and care in illustrating this book. This makes such a difference when dealing with a subject very much concerned with the appearance of objects. I am also indebted to Jane Paterson and Norma Hansell for additional last minute line drawings. I would also like to thank Liz Denton for her efficiency and good humour in the preparation of graphs and tables. My thanks also to Professor Charles Brown and Victor Scheffer for allowing me to reproduce their photographs. Finally, and most sincerely, I express my thanks to my editor, Tracey Sanderson, for her calm and patience.

CHAPTER I

Animal builders and the importance of bird nests

1.1 Introduction

Among the hundreds of bird nests carefully preserved in The Natural History Museum, London, is an unremarkable looking cup nest of grass and rootlets collected in New Zealand during the last century. It is the nest of the mournfully named *piopio or New Zealand thrush (*Turnagra capensis*). The interesting thing about the nest is that the builder is extinct, a victim of introduced mammals and its own over-confiding nature. The bird itself was last observed in 1947 (Fuller 1987). Possibly no other nest of this species remains in the world. It is an enduring expression of behaviour that can no longer be seen. To touch it is to be as close to its maker as to touch a brush stroke of a Van Gogh sunflower.

The structures animals build persist through time and in their construction the world is changed for the builders and for others. They may simply be regarded as objects of curiosity or wonder but, as products of evolution, they must embody principles of the organisation of the behaviour that created them, of functional design and of the evolutionary process itself; the scientific study of them should reveal some of these principles. This is the purpose of this book and the subjects to be examined are the nests and other objects constructed by that most consistently impressive group of animal builders, the birds.

So, in an important sense this is not a book about birds but about nests and related construction behaviour. Other groups of animals build nests or other structures, some much more remarkable than those built by birds. All these organisms share features in common with birds. They are equipped for building and have behaviour which ensures that the anatomy, even if unspecialised, is effectively employed on the chosen building materials. Consequently, the building activity of birds can be viewed as a model system, the study

---

* An asterisk is attached to any species referred to in this book whenever some aspect of its nest is referred to and that nest has been included in the museum nest survey, e.g. *piopio (*Turnagra capensis*).

of which allows generalisations to be made about all animal builders and all built structures. This chapter identifies important features of animal builders as a whole. In this way the relative importance of birds as builders can be judged and their ability to contribute to an understanding of building biology assessed.

1.2 Builders extend their control

The New Guinea ladder-web spider (Araneidae) builds a prey capture device which is like a conventional orb web except that the bottom sector of the web has been enormously extended downwards (Fig. 1.1). The length of the web is more than 130 times the body length of the spider, a huge extension of the capture range of the animal beyond the tips of its limbs. This extension of the animal also persists in time, although not for very long; built in the evening, it is dismantled before dawn. It is a design probably specialised for the capture of moths which, having hit the ladder, struggle to free themselves from its sticky threads by shedding their lose scales. In doing so they roll down the web and would fall off the bottom of a conventional orb web but in this case lose all their scales and are trapped as their supply of scales runs out (Robinson & Robinson 1972). Animals build artefacts to extend their control over the environment.

The highest mounds of some Australian termites of the genus *Amitermes* approach 7 m, with a similar girth near the base. This is large enough to contain a few million termites. When this is compared with the planned Millennium Tower, Tokyo, which will be about 800 m high yet accommodate only about 50 000 office workers, the scale of the termite building can be appreciated. The life of termites is, in fact, inseparable from the structures they build. A large proportion of colony members never venture out of the mound, which with maintenance may last decades, far longer than the individuals that built it. In some species the environment created inside the mound is regulated within narrow limits by its architecture. The mounds of so-called 'magnetic termites' (*Amitermes meridionalis* and *A. laurensis*) are laterally flattened near North–South axis, and taper to a narrow ridge at the top. This ensures that in the dry winter months the nest is warmed in the early morning by the sun striking its eastern face and its temperature is maintained into the late afternoon by the sun on its western face; at midday, however, the mound exposes little of its surface to the strong rays of the sun overhead. In this way the nest temperature remains quite constant during most of the day (Fig. 1.2, Grigg 1973).

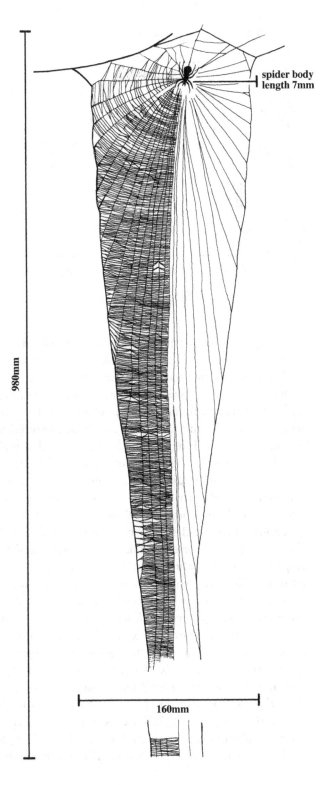

spider body
length 7mm

980mm

160mm

Figure 1.1
Builders extend their
control: the elongated
capture surface of the
New Guinea ladder-web
spider (Araneidae) is
about 130 times the body
length of the spider itself.
(Adapted from Robinson
& Robinson 1972.)

Figure 1.2
Temperature control through architecture: in its normal north–south orientation, the nest of the termite *Amitermes meridionalis* maintains a relatively constant temperature below 35 °C during daylight (solid circles). When the nest is rotated through 90°, nest temperature peaks at near 40 °C (open circles). (Adapted from Grigg 1973.)

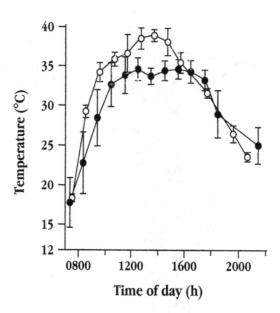

Heavy rains fall in the summer months, flooding the ground, but, with a height of 2–3 m, a mound of *A. meridionalis* keeps its one million or so inhabitants in safety well above the water level. As the water recedes the grass grows rapidly and the termites emerge to gather it into a multitude of storage chambers. This provides them with hay on which they can survive in the arid external world of the winter dry season. Temperature, humidity, food availability, these are all controlled by the agency of the mound itself. The termites' world loses its unpredictability, allowing them to transfer some functions formerly carried out by the bodies of their cockroach-like ancestors to the mound itself. Without the threat of desiccation or predation, their cuticles have become thin and soft. In their enclosed, darkened world, vision, body pigment and rapid escape responses have been almost lost. Birds' nests are by contrast designed to provide a controlled environment almost always simply to protect young. The few species in which adults sleep in nests outside the breeding season, even making dormitory nests (Skutch 1961), are the exception. The nests extend the control of parents in two respects: they help maintain the eggs near the typical avian body temperature of about 40 °C (Calder & King 1974), often considerably above the environmental temperature; and they reduce predation risk. A change in the structure of bird nests might therefore substantially alter the survival of eggs and chicks. This in turn might exert an indirect influence on the biology of the adults, for example the lifetime pattern of reproduction. This interaction between nest

and the organisms whose world it controls is one which runs
through this book.

1.3 The extended phenotype concept

The question of whether the artefact is in some respect part of the
organism that created it was explored by Dawkins (1982) in *The
Extended Phenotype*. He argues that, in the same way that there are
demonstrably genes for body form or eye colour, there must be
genes whose phenotypic expression is apparent in the architecture
of a nest or web. This concept does not seem to me to pose any
particular problems, although we know all too little about
phenotypic variation in animal-built structures or its genetic basis.
However, if we suppose that the number of radii in the orb web of
the garden spider, *Araneus diadematus*, has a genetic basis, then it
follows that, if web variants with more radii were more successful
than those with fewer, then spiders with the genotype for many
radii would increase in the population at the expense of those with
the phenotype for fewer radii.

The relationship between genotype and extended phenotype
may be more complex than this, and the evolutionary consequences difficult to predict. A nest builder may not affect its own
survival but that of its offspring, which will die if the nest falls out
of the tree. Nevertheless, better nest building will be selected for in
the same way as any other kind of parental behaviour since the
fitness of the builder is directly affected by the nest quality. Suppose, however, the nest is built by both parents – it is then an
expression of aspects of the genotype of both; or that it is built by
unknown thousands of individuals over many generations, as is a
termite mound. With several individuals responsible for the architecture, what then is the effect of selection pressures acting upon
the varied, mosaic nest phenotypes? Do they, for example, have
consequences for mate choice or rates of evolution? None of this is
known yet it is clear that, even in such complex systems of
phenotypic expression, the architecture can be sophisticated and
distinctive. The closed (Fig. 1.3) and open air conditioning systems
of *Macrotermes* mounds (Luscher 1961, Darlington 1984) apparently attest to this.

The garden spider builds an orb web and rests at its hub awaiting
the impact of flying prey. The related orb web spinner (*Zygiella
x-notata*) hides in vegetation at the side of the web, holding one
of the radii which connects directly to the web hub without contact with intersecting capture threads; this is a species-specific

Figure 1.3
Collective extended
phenotype: in the closed
ventilation system found
in some nests of the
termite genus
Macrotermes, the heat
generated by the colony
causes air to rise in the
nest core, driving it round
an air conditioning system
that regulates temperature
and gas exchange. The
mound is built and
extended by very large
numbers of termites over
many years. (Adapted
from Luscher 1961.)

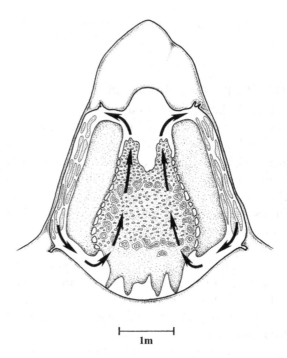

1m

difference. The architecture of an extended phenotype may be as
diagnostic of the species as the anatomy of its maker. Niche
specialisation is, in fact, frequently expressed through species-
distinctive architecture and may even be the feature which most
readily identifies species in the wild. Species recognised through
differences in their nest architecture have been labelled *ethospecies*
because they are recognised through their behaviour. Examples
have been described among wasps of the genus *Eustenogaster*
(Stenogastinae, Vespidae) (Sakagami & Yoshikawa 1968, Hansell
1984, Fig. 1.4) and termites of the genus *Apicotermes* (Schmidt
1955). Bird nests also may be readily identifiable to species; the
hanging nest of the *red-headed weaver (*Anaplectes rubriceps*,
Ploceinae), made of fine twigs held together by twists of dried bark
(see Fig. 9.2), is unique in construction.

Species-typical phenotypes may also be expressed with some
degree of variation. Whether natural selection acting on this can
produce evolutionary change depends upon whether the variation
has a genetic basis. Here our ignorance is strongly evident. There
are few systematic studies on the extent of variation in artefacts
built by any one species, fewer still on its cause or relative fitness of
the variants. The silk worm (*Hyalophora cecropia*), when it
pupates, constructs a cocoon which may be either *compact* or

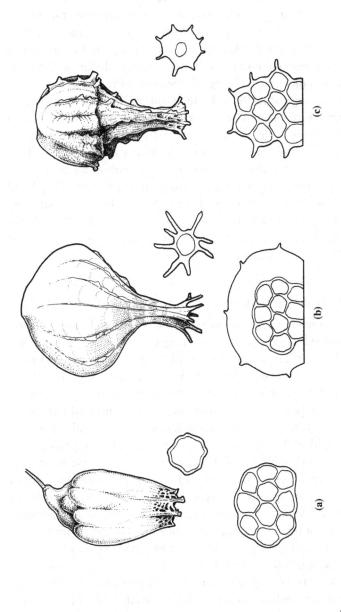

Figure 1.4
Ethospecies: the nests of three species of hover wasp (Stenogastrinae), each shown as side view, nest entrance shape and transverse section at the level of the brood comb. The nests are very distinctive, although it is difficult to distinguish the species on the basis of the wasps themselves. (Redrawn from Hansell 1984.)

baggy, but this polymorphism may not reflect genetic differences but be conditional upon the configuration of twigs where the caterpillar chooses to build (Waldbauer, Scarbrough & Sternburgh 1982). The stenogastine wasp *Parischnogaster mellyi* constructs a nest which may be a row of cells placed more or less end to end along a fine plant stem; cells may, however, be placed on the sides of previous cells to form a compact comb design or placed to generate an intermediate design. Again, the design appears to be conditional on environmental circumstances. The elongate design is preferred, more comb-like designs being adopted when the nest suspension is too short, suggesting that it might survive better, for example, through concealment from visually hunting predators (Hansell 1981). Within-species variation in nest design could alternatively be the result of differences in experience. Where construction behaviour is complex, individuals may improve with experience. Birds are potentially a very interesting group in this respect. They build rather elaborate structures and their capacity to learn is strongly evident in other aspects of their behaviour. In species in which this could be demonstrated, the prediction would be that experienced birds build better (more protective) nests.

1.4 Building behaviour changes habitats

By their behaviours, builders change the world and by their activity over succeeding generations, they may create distinctive new habitats, which represent a long history of occupation by the builders. In Cape Province, South Africa, the landscape is covered over large areas with more or less circular mounds 30 m across and 2 m high. These are apparently the consequence of the combined burrowing activities of termites (*Microhodotermes viator*) and mole rats (Bathyergidae) (Lovegrove 1991). This so-called *mima prairie* landscape is also found in the USA, created in this case by pocket gophers (*Thomomys*) (Fig. 1.5), and in Argentina, due to rodents of the genus *Ctenomys* (Cox 1984, Cox & Roig 1986). In South Africa the termite *Odontotermes* is thought to be responsible for creating a landscape of regular corrugations with parallel gullies about 50 m apart separated by ridges about 2 m in height which may be 1 km in length (Sattaur 1991). The period of time over which these landscapes have been shaped by burrowers is unknown, but the evenness of the mounds in *mima prairie* suggests that individual burrow systems may have been in more or less continuous occupation for decades and quite possibly a great deal longer (Hansell 1993a). An active mound of the termite *Macrotermes goliath* was found to

Figure 1.5
Builders change habitats:
mima prairie in North
America; a regular
landscape, shaped over
long periods of time by
the burrowing activity of
pocket gophers,
Thomomys. (Photograph
by V. Scheffer.)

contain archaeological remains dateable at 700 years old (Watson 1967).

The nests of birds, by contrast, are generally seasonal and short lived; the nest of a bird such as a *chaffinch (Fringilla coelebs) has a working life from first egg in to last chick out of no more than 35 days (Harrison 1975), although nests of large raptors may last for many years. The nests of megapodes are, in avian terms, exceptional. They may be used and added to year after year. In northern Australia there are sites where several hundred of these nest mounds occur together, some over 10 m high. As local megapodes weigh only about 1 kg, these mounds were originally identified as Aborigine middens (Bailey 1977) but now seem likely to be the nest mounds of the scrub fowl (*Megapodius (freycinet) reinwardt*) (Stone 1989). In New Caledonia mounds of 50 m diameter and 4–5 m high may be either archaeological sites or mounds of the extinct megapode *Sylviornis neocaledoniae* (Mourer-Chauviré & Poplin 1985).

1.5 Exploitation of the resources by others

The concentration of resources in a nest in the form of the occupants, their food supplies, the favourable microhabitat, even the nest material itself, creates a new range of niches to be exploited by a variety of specialists. The rich resources enclosed in social insect nests have led to the convergent evolution of a number of mammalian predators with long tongues and reduced dentition, including the aardvark (*Orycteropus*) in Africa and giant anteater (*Myrmecophaga*) in South America. Birds such as the honey buzzard (*Pernis apivorus*) and the red-throated caracara (*Daptrius americanus*) are specialist predators of the nests of wasps and bees (Thiollay 1991), as are some social insects themselves such as the tropical hornet (*Vespa tropica*) in Asia and the army ant (*Eciton hamatum*) in the New World tropics. A large number of arthropod species also exploit social insect colonies from within the nest itself in ways ranging from feeding on nest debris to obligate parasitism. Considering only ants, the arthropod symbionts found in their nests extend across six classes and ten insect orders which include 35 families of beetles alone (Hölldobler & Wilson 1990). Most obviously, however, the nest structure itself represents a resource which is of value to the conspecifics of the builder. The preferred method of nest foundation by the stenogastrine wasp *Eustenogaster calyptodoma* is by the acquisition of the nest made by another female (Hansell 1986). As rather ephemeral structures, bird nests provide fewer opportunities for exploitation, and the biodiversity of nest commensals is low. Nevertheless, chicks are a rich and vulnerable source of food, attracting predators or supporting a large population of nest parasites. This can have severe effects on chick survival or development, and has resulted in some construction behaviour and design features of nests to combat these problems (see Chapters 5 and 6).

1.6 Nests reinforce social life

Darwin, in *On the Origin of Species* (1859), admitted that he at first felt that the presence of sterile castes in the nests of social insects might be '*fatal*' to his theory of evolution by natural selection. He overcame the problem by suggesting that selection in social insects takes place at the level of the colony, allowing sterile morphs to be expressed in each generation. When, in 1964, Hamilton expounded his concept of *inclusive fitness* it could for the first time be clearly argued that individuals might benefit by sacrificing personal fitness to enhance the fitness of close relatives. Lin and Michener (1972) also argued that co-operative behaviour might evolve between non-relatives even if this *mutualistic* relationship favoured one party more than another, provided that both parties were better off co-operating than acting alone. These theories have no special associations with nest builders but do help to explain how highly co-operative nest building might have evolved, but is it possible that nest building itself might have contributed to the evolution of greater social complexity?

Nest building, by creating a more stable environment for the builders, could allow their population to rise close to the carrying capacity of the environment. Difficulty in founding a nest in this competitive environment might result in offspring delaying departure from the home nest and helping their parents in order to obtain some inclusive fitness. Larger colonies would then lead to further elaboration of the nest (Hansell 1987a). This can be described as a *lack of opportunity* hypothesis since this is what causes offspring to delay the onset of their own reproduction. Another nest-centred theory, however, predicts exactly the same outcome. This is a *cost of building* hypothesis which claims that it is the expense of founding the new nest that causes offspring to tarry in the home nest. No direct evidence is available to confirm either hypothesis, but examination of the occurrence of social insects suggests that a correlation between nest building and the evolution of social complexity does exist.

All termites (Isoptera) have advanced societies with sterile castes, as do all ants (Formicidae, Hymenoptera). Among the wasps, some are social and some are solitary, but the technological advance of light, strong, paper nests, the successor to burrows or mud nests, seems to have been accompanied by the evolution of larger colonies with predominantly sterile workers (Hansell 1987a). The ambrosia beetle (*Australoplatypus incompertus*), which digs deep burrows in trees that may remain occupied for more than 35 years, is exceptional among coleoptera in having a queen-like female accompanied by smaller workers. Naked mole rats (*Heterocephalus glaber*) live in colonies of 100 or so which contain only a single reproductive pair. The society is strongly reminiscent of termite colonies and very unlike that of more typical mammals. The species is also atypical in being completely obligate in its subterranean life, eating, sleeping, feeding and mating within the extensive system of burrows and galleries created by the workers.

So, whether through lack of opportunity or cost of foundation, it does seem that nest building has a self-reinforcing effect through evolution when it leads to delayed departure of the offspring of colony members. But the prediction must be that this will make a greater contribution to the evolution of social behaviour in species whose environment is substantially created by themselves rather than in species like birds which spend only a small proportion of their lives in nests and have no social structure comparable to that of social insects. Some, like certain species of jay (Brown 1974), do live in family groups partly created by the difficulty of young adults acquiring territories in a saturated habitat, but there is no evidence that the nest has contributed towards this situation.

1.7 The builders

The occurrence of building behaviour is neither confined to a narrow range of taxa nor scattered evenly through the animal kingdom. Instead, it has a few outbursts of virtuosity with talented displays of skill occurring sporadically across the animal spectrum. Nevertheless, the examples of diversity and complexity are found predominantly in only three classes, two of them arthropod and one of them vertebrate. They are the classes Arachnida (spiders and mites), Insecta (insects) and Aves (birds). Among the arachnids it is the spiders (order Araneida) which show the building skills. The New Guinea ladder-web spider (see Fig. 1.1, Robinson & Robinson 1972) exhibits one variant of the elegant orb web found in two spider families, Araneidae and Uloboridae. Spiders in other families build tangle or sheet webs and spin themselves shelters or retreats; most spiders envelop their eggs in silken cocoons, some of complex design. In the class Insecta, the greatest display of building behaviour is found in four orders, two of which are dependent upon silk for their success; they are the caddisflies (Trichoptera) and the butterflies and moths (Lepidoptera). They spin their silk only as larvae in order to create individual shelters to protect them during the pupal phase and, in the Trichoptera in particular, the larval phase as well. Before pupation, the larva of the silk moth *Antherea pernyi* constructs a silken stalk at the end of which it spins a bag to enclose itself, using the remainder of its silk to create a cocoon lining of different texture (Lounibos 1975). Other species pull out their own body hairs at pupation and bind them together with silk to produce shelters of intricate design (Hansell 1984). Caddis larvae, which are predominantly aquatic, make portable cases out of sand grains or vegetation pieces stuck together with silk. Some species spin silken nets which capture prey swept along in water currents. *Macronema transversum* combines a shelter and trap by making a sandgrain shelter with an inhalant and an exhalant funnel

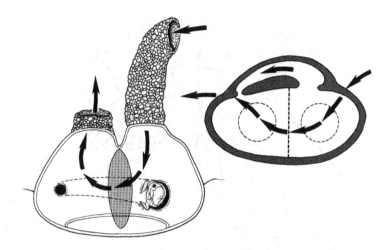

Figure 1.6
The case of the caddis
larva (*Macronema
transversum*) is a chamber
and side passage of sand
grains through which a
stream of water is
directed, entering by an
upstream funnel. Across
the chamber is fine silk
net which filters the fine
particles on which the
larva lives. (Adapted from
Wallace & Sherberger
1975.)

and spinning across the central chamber a fine silk web with a mesh
size of 3 × 30 μm (Wallace & Sherberger 1975, Fig. 1.6).

The two other insect orders surpass the Trichoptera and
Lepidoptera in their building abilities. They embrace nearly all
those species which are referred to as *social insects*. They are the
two quite unrelated orders Isoptera (termites) and Hymenoptera
(ants, bees and wasps). Examples of the size and complexity of
termite mounds have already been referred to, but colonies of ants
can be even larger. Those of the leaf-cutter ant species *Atta vollenweideri* may use 8 million *adults* in one colony (Fig. 1.7, Hölldobler
& Wilson 1990), a scale only approached by the largest human
conurbations. Their nests of subterranean burrows, chambers and
fungus gardens are, however, less impressive architecturally than
the arboreal nests of bees and wasps. In these, the importance of
self-secreted materials is again apparent. The critical material in bee
nest construction is wax; for wasps, which owe their architectural
success to the manufacture of paper, it is a salivary secretion to bind
together the plant fibres and larval silk to reinforce them which are
the essential ingredients (Hansell 1987a).

The third major class of animal builders is the birds. Among the
vertebrates, there are interesting nest builders among frogs and fish,
and some impressive burrowing mammals, particularly rodents.
The primates (monkeys and apes), on the other hand, are unimpres-
sive, with the extraordinary exception of ourselves. Birds, although
building abilities are not uniformly expressed among them, have
thousands of species that build a nest of some kind and hundreds
that build structures that are differentiated into discrete parts with
separate functional roles. Different parts of the nest are often characterised by the use of different materials and, across the class, at

Figure 1.7
A mature nest of the leaf-cutter ant (*Atta vollenweideri*) contains several million adults and would dwarf any person attempting to excavate it. (Adapted from Hölldobler & Wilson 1990.)

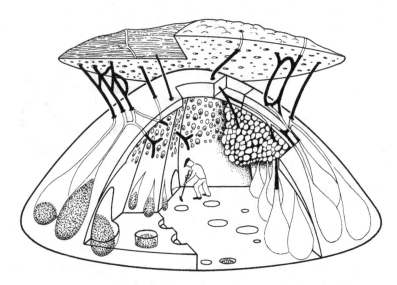

least as wide a range of materials is used as in any comparable invertebrate group. A few bird species also show other construction behaviour: i.e. the fabrication of tools and the creation of elaborate male display structures, bowers. Birds are therefore a particularly interesting group of builders to study, the more so because the behaviour of higher vertebrates is generally characterised by greater complexity and opportunity for modification through experience. Bird building behaviour therefore allows an assessment of whether an advanced nervous system offers any special advantages to builders and whether we can see in them aspects of building ability that are comparable to our own.

1.8 Are there shared characters among builders?

a) Simple minded

An important negative point to make about builders is that they do not need to be clever to be good at it, in fact it is not even necessary to have a nervous system. A number of classes of protozoa have quite impressive builders. The amoeboid organism *Difflugia coronata* lives in a portable sandgrain case only about 160 µm across which bears small spikes, also built of sand grains (Fig. 1.8, Ogden & Hedley 1980). The aperture through which the organism extends is ornamented with a delicate denticulate collar of fine particles. The organism reproduces by dividing into two, but before doing so picks up and stores internally the necessary quantity of the correct sized particles (Pateff 1926). At cell division, all these sand grains are somehow conveyed to the surface and arranged to provide a

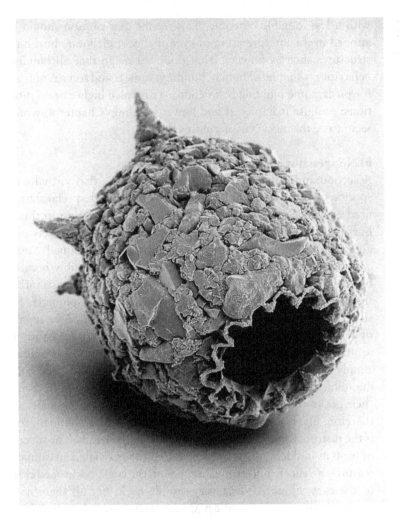

Figure 1.8
The intricate structure of
the minute sandgrain case
of the amoeba *Difflugia
coronata* demonstrates
that architectural
sophistication can be
shown by very simple
organisms. Magnification
approximately ×400.
(Photograph courtesy of
Natural History
Museum.)

new, completed case for a new daughter organism. All these building stages – selection, gathering and assembly – are in principle the
same as for a bird making a nest.

Structures which are regular and orderly may suggest some
sophistication in the organism which creates them, but I would call
this is a *crop circle fallacy*. Orderly outcomes can be the result of
very simple processes, even non-biological processes like the
growth of a crystal. Standard building units, a repetitious assembly
procedure and simple design rules produce orderly structures. A
honeybee comb is an impressive structure, but in foraging, bees
show navigation skills, topographical learning and improvement in
their handling of complex flower structures. Their ability to build is
no more surprising than their ability to perform other behaviours

with a less tangible outcome. This means that caution should be applied in the interpretation of even the most elaborate bird-built structures such as bowers. That does not mean that all building behaviour is simple; of human building, some is and some is not. So, for birds, some building behaviour may involve higher brain functions; tool use (Chapter 4) and bower building (Chapter 8) would seem to be the most obvious places to look.

b) No specialist anatomy

A second and rather surprising generalisation is that virtually no species of whatever builder possesses anatomical specialisations used exclusively for building. The interesting exception to this is burrowers, for example various species of moles, mole crickets, and an extinct Oligocene mammal, the appropriately named *Xenocranium* (Rose & Emry 1983, Fig. 1.9). This is possibly because burrowing equipment is used throughout life on a daily basis rather than sporadically or even occasionally. Spiders have spinnerets for silk production, but silk deployment is achieved by the legs, organs of general locomotion. Caddisflies have modified legs for handling sand grains, but these are also used in locomotion. Caterpillars simply make use of the general flexibility of their bodies. Wasps use mandibles to pulp prey as well as to prepare paper, and birds use their beaks and feet, organs with other obvious uses. Humans are therefore unusual in having special anatomy whose dominant role is the restructuring of the world about them, the hands. The success of birds as builders can be attributed to three associated anatomical features, a delicate but strong instrument, the beak, positioned close to the eyes, mounted on a very mobile neck, but all these have essential functions during the majority of the year, when there is no nest building.

c) Techniques shape materials but materials shape techniques

The environment offers only a limited number of types of building material and, regardless of the anatomy of the builders, there are only a limited number of techniques that can be applied to them. At its most basic there are only two techniques, *sculpting* and *assembly*, and only three materials *animal*, *vegetable* and *mineral*.

Sculpting is the removal of material to produce a desired shape. Almost invariably this is to create a burrow, for example the nesting burrows of motmots or jacamars (see Fig. 7.1). So, whether insect, bird or naked mole rat, burrowers chisel and scrape at the workface to loosen the substrate, creating spoil which must then be conveyed away. Assembly may embrace a larger range of techniques which

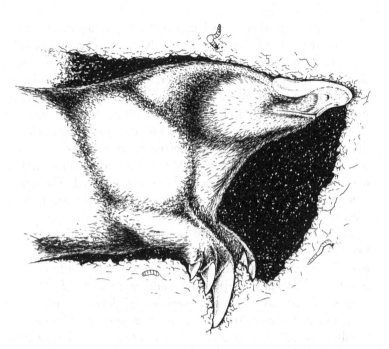

Figure 1.9
Reconstruction of the
Oligocene burrowing
mammal *Xenocranium*
showing specialised
digging anatomy of
massive claws and
flattened face. (Redrawn
from Rose & Emry
1983.)

not all animals will show, but there are universal elements of
gathering, transportation and *assembly*. If the artefact is itself portable, like a caddis case, then it is transported to the materials not the
materials to it. If it is fixed, like a nest, then other generalisations
also apply, these concern building costs, navigation ability,
specialisation in the selection of materials and choice of materials in
relation to the size of the organism.

Building has costs which can be measured as time and energy.
In either currency transportation is likely to be greater than either
gathering or assembly. For highly mobile organisms individual
journey times may be short, but their number may be large. Bearing in mind the additional load of the building material, this may
amount to a significant energetic cost (see Chapter 6). For slower
moving organisms, the time taken to gather the material may be
more important. In either case there will be a benefit in reducing
costs and this can be achieved by building the structure near to
the source of the material. Whatever the distance between materials and building site, the animal must be able to find its way
from one to the other and most nest builders will wish to return
to the nest repeatedly after its completion. Some navigation skills
and topographic memory abilities are a further expectation of
most builders.

It could also be argued that there is an advantage in being
versatile in the selection of materials so that an adequate supply

could be found in almost any location. There are, however, reasons why this might not be the case. If orderly structures can be built by applying simple building procedures to standard building units, then accepting a wider choice of materials will require more elaborate building performance. The loss of regularity in the structure might also cause structural weakness and, if a certain material is the best for the job, then any alternative would diminish the quality of the finished object. The examples given so far in this chapter support the conclusion that natural selection tends to favour specialists over generalists in the selection of building material. The occurrence and merits of nest material specialisation in birds are an often repeated theme in this book.

In practice, the most available source of materials is likely to be from plants since most animals are closely associated with them. Also, in terrestrial habitats at least, plants are quite differentiated structures and so produce potential building materials of a range of shapes and sizes (roots, stems, twigs, leaves, flowers, fruit) with a range of physical properties, so suiting the assembly techniques and building specifications of a wide variety of species. Some of these plant materials may be found littering the habitat in more or less standard units. Others, like leaves, have the merit that standard units may be cut from them, like the leaf panels in the case of the caddis larva *Lepidostoma hirtum* which are cut and assembled to create the box girder inside which the larva lives (Fig. 1.10, Hansell 1972). The scale of the building units will also be appropriate to the size of the organism. A *woodpigeon, *Columba palumbus*, selects twigs about 20 cm long, each weighing 6–7 g (Hansell & Aitken 1977), less than 1.5% of its 490 g body weight. Wasps make nests of paper in which plant fibres no more than a few hundred microns long are bound together in a salivary matrix (Hansell 1987a).

If particular materials by their nature offer common potential and constraints to all builders, so structures of similar scale should be able to benefit from the same materials regardless of the organisms building them. The nests of some social wasps, for example *Apoica* and *Protopolybia* species, are made from hairs harvested from the surface of plant leaves or stems. The nests of the *black-chinned hummingbird (*Archilocus (colubris) alexandri*) and the *magnificent hummingbird (*Eugenes fulgens*) are also made of plant hairs (Fig. 1.11), and are of a size and weight comparable to those of a wasp colony (weight of nest + incubating bird + two eggs for the two hummingbird species is respectively 12.6 g and 6.0 g) (Hansell 1996a). Mud is an interesting exception to the general rule

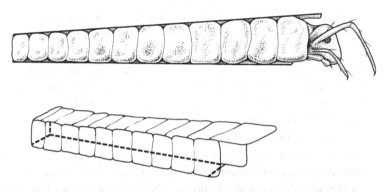

Figure 1.10
Standard building units:
the caddis larva
(*Lepidostoma hirtum*)
makes a square-sectioned
case by silking together
more or less rectangular
leaf panels in four rows.
(Redrawn from Hansell
1972.)

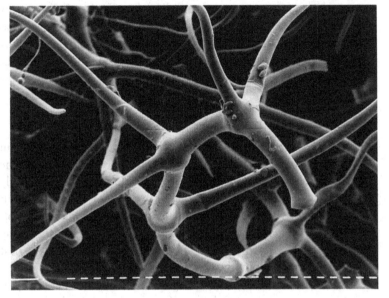

Figure 1.11
Materials appropriate to
the scale of the structure:
the nest of the
*magnificent
hummingbird (*Eugenes
fulgens*) is made of plant
hairs. (Scale: separation
of two white bars is
10μm.) (Photograph by
Margaret Mullin.)

that particular materials are appropriate to the building of structures of a certain size. It is used to make the brood cells of eumenid
wasps (Vespidae), the nests of several species of birds, and, in Mali,
palaces for humans (Armstrong 1988).

Preventing a structure from falling apart is a problem universal
to species that create structures by assembly. It is not such a serious
problem in structures built from the ground up, because the weight
of the top materials can contribute to the whole structure becoming
more compact. But for structures built from the top down there are
essentially only two ways that they can be prevented from tearing
themselves apart. They must be *bound* together or they must be
stuck together. It is a problem that must be solved by all organisms
that build structures in tension and it is instructive to notice that the
major groups of invertebrate builders have solved it with secretions

from their own bodies, e.g. silk (spiders, lepidoptera, larvae such as silk worms), wax (social bees), faecal cement (termites).

Self-secreted materials are not always glues or even plastic materials, they may be discrete, ready-made building units, for example the down feathers used by the common eider duck (*Somateria molissima*) to line her nest or the rod-shaped faecal pellets excreted by the caterpillar of an Australian moth, which are stuck together with the larva's own silk to make a space-frame subsequently covered with silk to create a retreat (Clyne, personal communication). Some building materials of animal origin are created by species other than the builders. The *burrowing owl (*Speotyto cunicularia*), for unknown reasons, lines its nest chamber with cow or horse dung. Good building materials secreted by one species, as will be very apparent in Chapters 4 and 5, may also be recycled by other species to create quite different structures. In the case of birds, that material has been pre-eminently arthropod silk.

1.9 Chapter by chapter

Bird nests are virtually all an extension of parental care, although in a few species, nests may be built and used as dormitories outside the breeding season (Skutch 1961). Bird reproduction cannot be fully understood without consideration of the nest, nor the nest without reference to its intended contents. Chapter 2 examines the evolutionary dialogue between the demands that the eggs and the young exert upon the design of the nest and the constraints that the nest itself may impose upon the clutch.

As this is a book about building behaviour, it seeks not only to provide an assessment of the significance of nests in the breeding biology of birds, but also to use nests to examine and develop some generalisations about animal building behaviour as a whole. To do this it was necessary to have descriptions of nests that were obtained from a common set of criteria. This required the development of the nest profile check sheet which is described in Chapter 3. It was developed for obtaining data by non-destructive examination of nests in major museum collections in the UK, France and USA. The nests of 518 different species were examined using the profiling scheme. Apart from a handful of cases, each nest was of a different genus. These represented 145 non-passerine genera (i.e. 16% of non-passerine genera taken from 45% (44) of the families), together with 373 passerine genera (i.e. 32% of passerine genera taken from 71% (32) of the families). In consequence, the nests examined represent a good sample of the full range built by birds.

Making a nest is making bits of material stay together in a certain spatial relationship. Birds collectively make use of a wide variety of building materials. Materials influence construction behaviour and, in some measure, also shape the architecture of the nest itself. The varied construction principles used in bird nest building are outlined in Chapter 4, which also provides an opportunity to assess whether construction should be regarded as complex. This is an important issue since there is an ambivalence of attitude here that, on the one hand, regards nests as remarkable structures, while, on the other, categorises the building behaviour as largely genetically determined and inflexible. The issue of the complexity of nest construction behaviour is dealt with in Chapter 4.

Nests are generally differentiated structures with different parts performing different functions. The data gathered from the museum nest survey are analysed in Chapter 5. The characteristic materials of the structural and lining layers of nests, the outer decoration and attachment devices of the nests examined in the survey are described and the relationship between aspects of nest design considered.

The reproductive cycle in birds entails various costs: those of egg production, incubation, chick rearing and the cost of building the nest itself. Until recently, nest building costs had been ignored or assumed to be slight, but, with a growing interest in the importance of reproductive costs in shaping the pattern of lifetime reproduction, an accurate assessment of nest building costs is now needed. Chapter 6 reviews the available evidence on nest building costs.

The success of a nest in carrying out its task lies not only in the quality of the construction, but also in where it is located. This is in fact the aspect of bird nest biology for which we have most extensive evidence. The siting of the nest in response to the physical and biological hazards that confront the clutch is the subject of Chapter 7.

Less than 20 species of birds out of a total of more than 9500 are celebrated for the construction not of nests but of bowers, structures produced by males to attract females. The whole of Chapter 8 is devoted to considering the importance of these structures because they have now become a model system for the investigation of sexual selection through female choice. The males of many animal species exhibit features of anatomy and behaviour shaped by sexual selection, but bowers are unique in being display objects yet separate from the male whose reproductive success depends upon it; they, the bowers, are therefore particularly amenable to experimental manipulation, and also expressed completely through the male's behaviour rather than his morphology and behaviour.

Bird nests are varied in composition and in architecture. Examination of the nests of closely related species exposes the history of the change of nest building in that group. This reveals the flexibility of building behaviour through evolution and allows the importance of changes of building behaviour in promoting speciation to be considered. These particular and more general aspects of nest evolution are the subject of Chapter 9.

1.10 The taxonomic convention

The classification of birds and species nomenclature used in this book are taken from Sibley and Monroe (1990). This classification is derived from DNA-hybridisation studies (Sibley, Ahlquist & Monroe 1988). This system is not to everyone's liking, but has the merit of being comprehensive and based on a uniform criterion. It recognises 9672 species in 2057 genera derived from 21 orders. By far the largest order is the Passeriformes (Passerines), with 5712 species in 1161 genera. It is this one order that contains the bulk of diversity in nest construction and architecture, and inevitably provides the great majority of examples used in this book.

Sibley and Monroe (1990) separate the Passeriformes into two sub-orders, Tyranni (291 genera) and Passeri (870 genera). This is not particularly important in the context of this book; however, it does concur with the more traditional separation of the order into the respectively Sub-oscine and Oscine Passeriformes. Overall, the classification of Sibley et al. (1988) and Sibley and Monroe (1990) matches quite closely that followed in more traditional taxonomic schemes; however, it differs markedly from them in recognising only 46 passerine families compared with a more traditional scheme, for example Howard and Moore (1991) which recognises 74.

The most notable families among the small number in the Tyranni are the tyrant flycatchers (Tyrannidae: 146 genera), antbirds (Thamnophilidae: 45 genera) and ovenbirds (Furnariidae: 66 genera), but also included are the pittas (Pittidae) and broadbills (Eurylaimidae). The most prominent families in the sub-order Passeri are the honeyeaters (Meliphagidae: 42 genera), the diverse family Corvidae of 127 genera ranging form large birds like crows to small species such as monarchs, the thrushes and chats (Musci-capidae: 69 genera), the warblers (Sylviidae: 101 genera), the sparrows and estrildine finches (Passeridae: 57 genera) and the buntings, cardueline finches, tanagers, etc. (Fringillidae: 240 genera).

CHAPTER 2

The clutch–nest relationship

2.1 Introduction

Bird reproduction has some distinctive and unusual features compared with other vertebrates. All birds lay eggs; there is not a viviparous species among them (Blackburn & Evans 1986). The eggs are also remarkably large (10–20% of body mass compared with < 3% for modern reptiles) (Wesolowski 1994). About 90% of bird species show biparental care, although this is rare in other vertebrates. Parental care in most bird species also includes the construction of a distinctive nest and the incubation of the eggs (Wesolowski 1994). How these associated features of bird reproduction came about and the role of the nest in that association are the subjects of this chapter.

2.2 The nests and brood care of dinosaurs

In 1923, members of the American Museum of Natural History expedition to Mongolia discovered for the first time whole clutches of eggs laid by dinosaurs (Andrews 1932). These eggs were from the Cretaceous but clutches from the Lower Jurassic and even Upper Triassic, in what appear to be prepared nest sites, have now been described (Moratalla & Powell 1990), demonstrating that land vertebrates were constructing some sort of nest as early as 200 million years ago. Dinosaur nests are now known from sites around the world, mostly dating from the Upper Cretaceous (about 80 million years ago), but the pattern of egg laying in dinosaurs is still only known for a handful of the roughly 285 genera (Moratalla & Powell 1990).

Two major orders of dinosaurs are recognised: the order Saurischia contains large carnivorous species like the 6–9 m Tyrannosaurids and also the massive (27 m) Diplodocids, while the order Ornithischia contains large herbivores such as *Triceratops*, *Stegosaurus*, duck-billed hadrosaurids and the relatively small (2–5 m) hypsilophodontids and parrot-beaked protoceratopids. Evidence now seems to confirm the origin of birds from a subfamily of the Saurischia, the Theropoda (Padian 1998). Two genera of feathered but flightless theropods, *Protarchaeopteryx* and *Maniraptora* of Late Jurassic or Early Cretaceous date (about 145 million years BP), have recently been found in China (Quiang *et al.* 1998), establishing a link between these dinosaurs and the earliest known bird, *Archaeopteryx*, which dates from the Late Jurassic.

The creation of prepared nest sites is recorded for both Saurischia and Ornithischia. The Cretaceous dinosaurs show two distinct nesting habits, each associated with a particular egg shape. Near-spherical eggs are laid in excavated pits, whereas more elongated eggs are placed, with their long axes vertical, in prepared mounds (Fig. 2.1). Clutches of 30 or so spherical eggs of a probably sauropod dinosaur (*Faveoloolithus*) arranged in two or three layers have been found in Mongolia. The gravelly substrate in which they are laid suggests that the nest pits were dug on a sand bar at the end of the rainy season and the eggs deposited in two or three stages, with the sand pushed back into the hole to cover each layer (Mikhailov, Sabath & Kurzanov 1990). These Mongolian deposits also reveal nests of ornithischian dinosaurs *Protocertops* and *Breviceratops*, in which clutches of about 20 elongate eggs have been laid together, apparently in the soft sand of a prepared mound, and then covered (Mikhailov *et al.* 1990).

Both these types of nests are also constructed by living reptiles. The green sea turtle (*Chelonia mydas*) is typical of the chelonians in digging a pit on the beach in which the eggs are deposited, then covered and abandoned. Eight of the 21 surviving species of crocodile are also hole nesters; they probably represent the ancestral nesting behaviour of a group which, over a 200 million-year period, first accompanied and then survived the dinosaurs. Among them arose crocodile species that prepare mounds, predominantly of vegetation, in which the eggs are laid and then covered. This nesting habit is shown by four of the living 11 *Crocodylus* species plus the alligators and caimans (Greer 1970, 1971). Most of these living crocodilians, unlike the chelonians, do not abandon the nest after the eggs are laid, but guard it for as much as two or three months, finally helping to release the calling hatchlings from the mound and guiding them to water (Greer 1971).

Parental care in living reptiles is rare and performed by females, whereas in the most primitive living birds, the Eoaves (ratites and tinamous), parental care is almost universally male. Wesolowski (1994) concludes from this that ancestral birds showed no parental care and evolved first male then biparental care. This came about because the egg at first became larger under selection for greater precociality on hatching. Parental care then evolved to protect these

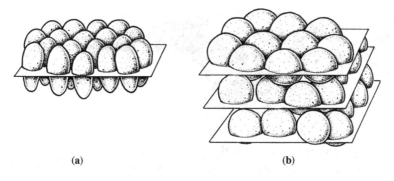

(a)   (b)

Figure 2.1
Clutch patterns typical of Cretaceous dinosaurs: (a) vertically oriented, elongated eggs laid in a mound by a species of Protoceratopsid; (b) spherical eggs layered in an excavated pit, e.g. in three layers by *Faveoloolithus*. (Redrawn from Mikhailov, Sabath & Kurzanov 1990.)

large eggs from predators. This role was adopted by males, who were already defending nest territories from rival males, while females concentrated on the production of more eggs. Incubation then evolved, according to Wesolowski, as a device to speed up embryonic development, biparental care being advantageous in this and for chick survival. Wesolowski (1994) identifies egg incubation, which brought about a marked raising of the temperature of embryonic development, as an important landmark in this evolutionary sequence, after which eggs and parent were necessarily coadapted. To which I would add that the nest also became a key part of that coadaptation.

There are, however, two particular problems with Wesolowski's (1994) argument of egg enlargement before the appearance of parental care. It assumes that superprecocial birds (i.e. megapodes), where there is no parental care, exhibit the primitive condition. Starck (1993), by contrast, regards their lack of parental care as secondary (see Section 2.4). In addition, Wesolowski (1994) makes little of the growing evidence of parental care in dinosaurs.

Some Cretaceous dinosaurs showed parental care beyond simply the preparation of a nest. Evidence of this has been found in ornithischian species from Upper Cretaceous sites in Montana, USA. Groups of nests of the same species, interpreted as nesting colonies, have been found containing eggs of a hypsilophodontid species (Horner 1982). Embryonic bone fragments from eggs identify these as nests of *Orodromeus makelai* (Horner & Weishampel 1988). In this species, clutches are of up to 20 or so vertically oriented eggs. The embryos of *O. makelai* reveal that limb bones were precocially developed, indicating that, like turtles or crocodiles, the hatchlings left the nest soon after emergence. However, the presence in the nests of the hadrosaur *Maisaura peeblesorum* of the bones of juveniles with teeth showing signs of wear, suggests that in this species hatchlings showed altricial development and

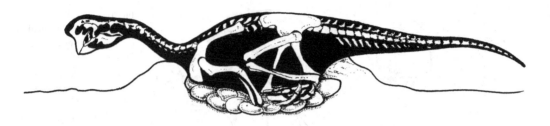

Figure 2.2
Reconstruction of an
Oviraptor fossilised in
egg-brooding position
after an apparently
sudden death. (Redrawn
from Norell *et al.* 1995.)

required parental care in the nest before becoming mobile (Horner 1982, Horner 1984, Horner & Weishampel 1988, Moratalla & Powell 1990). More exciting from the point of view of the evolution of parental care in birds is the discovery in Late Cretaceous deposits in the Gobi Desert of a small theropod of the genus *Oviraptor* fossilised in a brooding position over a clutch of about 20 eggs in a prepared nest site (Fig. 2.2), offering evidence that brooding behaviour evolved long before the origin of modern birds (Norell *et al.* 1995), although whether incubation of the eggs was involved in this case is very uncertain.

Direct evidence of parental care extending beyond nest construction in ancestral birds is at present lacking; however, the Mongolian Cretaceous sites do contain eggs of flighted Gobipterygiformes. The abundance of the nests of these birds along the ancient lake and river margins indicates that these birds probably nested colonially, while the good preservation of the eggs suggests that parents had buried them (Mikhailov *et al.* 1990). Evidence therefore points to a system of parental care in the earliest birds that at least entailed the preparation of a nest to contain the eggs. But what purpose did it serve? The nests of living birds provide the eggs and, in many species, the chicks also with protection from the extremes of climate and the threat of predation. The nests of chelonians and crocodilians, however, have another function; the determination of the sex of the offspring.

In the mounds of the American alligator (*Alligator mississippiensis*), eggs at 30 °C all develop into females, while those at 33 °C all become males (Deeming & Ferguson 1989). The emerging sex ratio results from the choice of nest site by the female and differences in temperatures in different parts of the nest. Whether dinosaurs used nests in this way is unclear. The concentric arrangement of eggs in clutches of *Protoceratops* might have created differences in incubation temperature within the nest that had significance for sex determination (Moratalla & Powell 1990). By contrast, it has been proposed that genetic not environmental sex determination may have contributed to the demise of the dinosaurs at the end of the

Cretaceous (Deeming & Ferguson 1989). The argument here is that this mechanism left them unable to evolve rapidly in a changing world, while crocodilians, in which sex was under a measure of environmental control, survived.

The birds, it is proposed, survived the end of the Cretaceous for another reason: they took to incubating the eggs in the nest with their own body heat, a situation which demanded genetic control of sex determination; and mammals survived in part by protecting early embryos within the female herself and giving birth to live young (Deeming & Ferguson 1989), a solution which has led to a vastly less interesting range of mammal nests compared with those of birds.

2.3 Why do birds lay eggs?

Not a single species of bird is known that does not lay an egg. This is rather surprising since not all fish lay eggs, amphibians and reptiles show both oviparity and viviparity and even the mammals, although largely viviparous, have some egg-laying species. The vertebrates in general seem to have been highly prone to the evolution of viviparity, which may have arisen independently as many as 118 times (Blackburn & Evans 1986). This has led to the proposal that one or more constraints may have prevented the evolution of viviparity in birds. Most obvious among these is flight, yet bats have successfully combined flying with bearing live young. Other suggestions include the necessity of chicks to break into the air space in the egg a few days before hatching or that the water-conserving, shelled (cleidoic) egg, if retained in a uterine environment, would create too great a barrier for the gas exchange needs of a developing chick (Blackburn & Evans 1986), or that a physiological constraint exists that prevents viviparity from evolving in any vertebrate species with a body temperature of 40 °C or above, thus excluding the possibility in birds (Anderson, Stoyan & Ricklefs 1987).

Blackburn and Evans (1986) attribute the absence of viviparity in birds to a combination of factors, in particular loss of fecundity due to limited space in the female reproductive tract and loss of opportunity for involving males in parental care. With viviparity, the mother provides a continuous supply of nutrients and water to the developing embryos, and disposes of their nitrogenous waste. In addition, the mother's body offers insulation and other protection from the hostile world outside. Dunbrack and Ramsay (1989) claim that a more important constraint on the evolution of avian viviparity has been the high oxygen demand of large eggs retained in the uterus of an endothermic animal. They point out that in mammals viviparity has been achieved through great reduction in egg size and the birth of highly altricial young fed on the uniquely mammalian diet of milk, possibly an evolutionary constraint in birds. However, as Blackburn and Evans (1986) point out, the cleidoic egg, a uniquely avian innovation, solved the problems of nutrition and waste disposal; the problem of protection from an unpredictable world, I would suggest, has been overcome through the great elaboration of the nest in birds compared with mammals. With that also came an opportunity for male birds to show a much greater degree of parental care than in mammals (Ligon 1993), through nest construction, incubation and brood care.

It does therefore seem quite likely that the nest of birds has been a significant element in their rejection of viviparity. Nests change the world, allowing builders to define part of the environment in which they live (Hansell 1987a, 1993a, 1996a). Dinosaur fossil clutches give support to the view that nests were a feature of bird reproductive biology from the outset. Subsequent evolution has refined the relationship between eggs and nest to an adaptive peak from which other successful solutions may be hard to reach through natural selection.

2.4 Do chicks need nests?

In the radiation of birds there has been a diversification in range of nest sites, in clutch size and in the dependence of the young upon the nest after hatching. Nest sites and nest designs in relation to nest sites are considered in Chapters 5 and 7, but to understand nest designs it is also necessary to consider what has led to the chicks of some species emerging from the egg helpless and blind while those of others leave the nest within minutes, and why species-typical clutch sizes range from one to over 20.

Starck (1993) recognises eight degrees of development that may be shown by chicks at the time they emerge from the egg (Table 2.1). These range from chicks hatching covered with down (even feathers), eyes open and able to feed themselves, to those that hatch blind, naked and helpless (Fig. 2.3). Chicks of *superprecocial* species (Megapodiidae) are all incubated in mounds whose temperature is generated by fermenting vegetation or heat from the sun and regulated through the behaviour of the attendant male. The megapodes are a monophyletic group of galliform birds whose ancestors were therefore *precocial* species resembling those of other galliforms such as pheasants. These typically build ground nests

Table 2.1. **Eight classes of chick development from (1) extreme independence to (8) total dependence shown by different groups of birds**

1. Superprecocial	No parental care, prolonged incubation time and are able to fly soon after hatching (e.g. Megapodiidae)
2. Precocial 1	Follow parents and feed alone, downy feathered (e.g. Anatidae, Tinamidae, Jacanidae, some Phasianidae)
3. Precocial 2	Like precocial 1 but food is shown by parents (e.g. *Meleagris*, wild turkey)
4. Precocial 3	Like precocial 2 but young are fed by parents (e.g. Cracidae, Turnicidae, Gruidae, Rallidae)
5. Semiprecocial	Downy-feathered hatchlings leave the nest only in case of danger, then return; they are fed and cared for by the parents (e.g. Laridae, Stercorariidae)
6. Semialtricial	Downy-feathered hatchlings, fed by the parents and do not leave the nest; high postnatal growth rate (e.g. Ciconiidae, Ardeidae)
7. Altricial 1	Display no motor activity, downy feathered, eyes closed at hatching; high growth rate (e.g. Strigidae, Columbidae)
8. Altricial 2	Hatch without external feathers, eyes and ears closed at hatching, grow very rapidly (exception large sea-birds) (e.g. Passeriformes, some Psittacidae, Pelicaniformes)

Adapted from Starck (1993).

with eggs incubated by a parent. Mound building in megapodes is therefore an apomorphic character, not one homologous with that of some Cretaceous ancestor.

The ancestral condition for modern birds was, Starck (1993) believes, precociality with different degrees of altricial development, and hence dependence of the chicks upon the nest, evolving from it. However, altricial development appears to have evolved more than once and precociality may also have had several independent evolutionary origins (Starck & Ricklefs 1994).

Precocial species generally have relatively large eggs compared with altricial species and the proportion of yolk in them is also higher than in those of altricial species (precocial species yolk content 37%; altricial species yolk content 22%) (Starck 1993). These egg differences are associated with longer egg incubation

(a)   (b)

Figure 2.3
A contrast in (a) the
newly hatched chick of
the barred buttonquail
(*Turnix suscitator*)
(*Precocial 3*) with (b) the
helpless hatchling of the
Java sparrow, *Padda
oryzivora* (*Altricial 2*).
(See Table 2.1 for
definitions of stages.)
(Redrawn from Starck
1993.)

times for eggs containing precocially developed chicks. The large
yolks of megapode eggs are therefore an essential adaptation allowing the chick to hatch in a sufficiently developed state to dig its way
out of the nest mound and almost immediately fly away (Vleck,
Vleck & Seymour 1984). Hatchling brush turkeys (*Alectura
lathami*, Megapodiidae) resemble in terms of development a two-
week-old chick of a peafowl, *Pavo cristatus* (Phasianidae) (Starck
& Sutter 1994).

The great majority of birds have evolved altricial chicks. More
than 70% of all species are placed in the category *Altricial 2* (Starck
1993). This includes more than 5000 species of passeriformes and
around 1500 species of non-passerine families. It is among these
that the most complex nests in terms of building materials and
design have evolved. This has ensured that chicks are placed in
locations that are less accessible or less visible to predators or
protected from climatic extremes in an environment in which temperature is regulated. So, the evolution of more protective nests
provided an opportunity for the evolution of chicks that were
naked, blind and helpless.

Nakedness is a typical characteristic in emerging chicks of
altricial species and the adaptive significance of it appears to be
twofold. Firstly, it gives the chicks direct contact with heat provided
by the parent while they are still ectothermic; secondly, loss of body
temperature while not attended by a parent may reduce energetic
costs and consequently the cost of rearing chicks. In the cactus wren
(*Campylorhynchus brunneicapillus*), homeothermy develops grad-
ually during the chick phase, with thermoregulation absent until
nine days of age. Ricklefs and Hainsworth (1968) argue that the
delay in the growth of insulating down and of thermogenesis may

result in the allocation of additional resources to growth rather than maintenance – speeding up chick development and hence reducing the period of dependency in the nest.

2.5 Clutch size

Lack (1947) observed that clutch size is 'usually far below the potential limit of egg production', and concluded '... in nidicolous (i.e. altricial) species, the average clutch size is ultimately determined by the average maximum number of young which the parents can successfully raise in the region and at the season in operation'. So Lack hypothesised that food supply for the chicks was the overriding influence on clutch size; consequently, the larger clutches that are characteristic of species in more northerly latitudes compared with those in the tropics could be explained as due to the longer hours available for feeding. Populations were subsequently regulated, according to this hypothesis, by density-dependent predation, but that this had no influence on clutch size. Evidence that clutch sizes are smaller in the tropics compared with temperate regions has been provided in a number of studies. Moreau (1944), in a major comparative study on European and Equatorial African species, found that clutch sizes in passerine and non-passerine species within 5 degrees of the Equator were smaller than those in the more temperate latitudes (25° S) of South Africa, which were in turn smaller than those in related species in Britain, and concluded that clutch size is adjusted to population life expectancy. Snow and Snow (1963) showed that clutches of three Trinidadian thrush (*Turdus*) species were smaller than those of European species of the same genus. Skutch (1985) and Kulesza (1990), in extensive surveys of New World tropical and temperate species, found the same trend.

Skutch (1985), however, observed that, contrary to the prediction of the Lack (1947) food supply explanation, clutches raised by both parents were no larger than those raised by a single parent. He proposed an alternative hypothesis for the difference in clutch size between temperate and tropical species: predation. He found that nest failure in the wet tropics was around 66%, compared to 50% for temperate regions, figures broadly in agreement with the data compiled by Ricklefs (1969) on nesting success, although the latter refers to all causes of nest failure, not simply predation. Skutch (1985) proposes two possible advantages of small clutch size as an adaptation to high nest predation: firstly, that restraint in any breeding attempt leaves energy to engage in another in the event of loss, and, secondly, that small clutch size necessitates fewer parental visits during chick rearing and so a smaller chance of the discovery of the nest by predators. The hypothesis that activity at the nest increases predation was tested in observations on nests of the neotropical western slaty antshrike (*Thamnophilus atrinucha*, Thamnophilidae). Not surprisingly, adult arrival and departure were greatest during the nestling phase; however, nest predation occurred equally during incubation and chick rearing (Roper & Goldstein 1997). In a review of literature on nest predation in relation to nest site, Martin (1993a) found, in woodland sites, lower rates of predation on the ground layer compared with off-ground nests (see Fig. 7.7). Martin (1988a, 1993a) points out that evidence of larger clutch sizes and longer nestling periods in ground-nesting compared with off-ground-nesting species supports the role of predation in the determination of clutch size.

The effect of clutch size on the ability to produce a replacement clutch was demonstrated by Slagsvold (1984). He subjected two groups of great tits (*Parus major*) to clutch removal at 15 days. One group was manipulated to lay large clutches (ten to 11 eggs) and the other to lay smaller clutches (three to four eggs) before egg removal. Re-nesting, small clutch pairs re-laid quicker and fledged more chicks than the large clutch group, supporting Skutch's prediction that high incidence of nest failure will select for smaller clutches.

A stated assumption of the Lack (1947) hypothesis is that birds are trying to maximise reproduction for that season. The growth of life history theory has brought with it a realisation that during a lifetime of reproduction, there may be trade-offs between current and later reproductive success. For birds, where several years of reproduction are the norm, this is an important consideration. Charnov and Krebs (1974) introduced the concept of *optimal clutch size*, and describe a model designed to incorporate the possible positive relationship between clutch size and adult mortality. That relationship is now supported by varied evidence (reviewed by Nur 1988), for example that female house martins (*Delichon urbica*) that raise two broods in one year have lower out-of-season survival compared with single-brooding females (Bryant 1979).

The empirical test of the cost of reproduction is by clutch manipulation. The predictions here are, for the *trade-off* hypothesis, that clutch enlargement may lead to a greater reproductive success, but that this will be at the cost of lowered future survival or fecundity. By contrast, the *individual optimisation* hypothesis predicts that any clutch manipulation, enlargement or reduction, will lead to reduced fledging recruitment in that season (Pettifor, Perrins &

McCleery 1988). The evidence of over 40 brood enlargement studies of free-living species is examined by Vander Werf (1992). This reveals that the effect of brood enlargement is generally to produce more fledglings than do unmanipulated clutches. This is contrary to the individual optimisation hypothesis and to Lack's (1947) prediction that clutch size is limited by food availability. But what of the evidence that parents with enlarged clutches pay a price in subsequent reduced fitness?

In part because Lack (1947) identified chick rearing as the key cost to parents of an enlarged clutch, many brood manipulation studies have consisted of adding a newly emerged chick to the clutch. These have not always shown trade-offs (Pettifor 1993a, 1993b, reviewed in Lindén and Møller 1989). However, as emphasised by Monaghan and Náger (1997), manipulating the chick number alone neglects the possible cost of laying additional eggs and of incubating them.

The costs of nest building and their consequences are dealt with in Chapter 6. The costs of different stages of reproduction on subsequent reproductive success have now all been investigated in empirical studies (Lindén & Møller 1989). Artificial enlargement or reduction of nestling numbers of the collared flycatcher (*Ficedula albicollis*) showed a correlation between manipulated clutch size and reduction of female fertility the following season (Gustaffson & Sutherland 1988). This study therefore identifies a cost in chick rearing. For territorial species, this cost might depend upon the quality of the territory. Brood manipulation in black-billed magpies (*Pica pica*) involving adding or removing a newly hatched chick showed that magpie parents did lay a clutch adjusted to territory quality (Högstedt 1980). This could explain some within-species variability in clutch size.

Costs have also been identified in egg production that could influence clutch size. Lesser black-backed gulls (*Larus fuscus*), induced to lay an extra egg by the removal of the first, hatched a lighter chick from the replacement egg whose chance of survival to fledging was only 25% that of the last egg in the clutch of the control group (Monaghan, Bolton & Houston 1995). Extra costs at the egg stage may not simply be in egg production but in incubation. In the common eider duck (*Somateria mollissima*), the costs of incubation can be observed in the positive correlation between the decline of body condition and clutch size during the incubation phase (Fig. 2.4, Erikstad & Tveraa 1995). This suggests that optimum clutch size in this species is determined by an interaction between the allocation of body reserves to egg production and incubation.

Figure 2.4
The relationship in the common eider duck (*Somateria mollisima*) between body condition of females and the size of their clutches showing: (a) a negative effect of clutch size on body condition at day 20 of incubation; (b) the increase in the relative loss of body mass with clutch size. Numbers indicate sample sizes (Erikstad & Tveraa 1995).

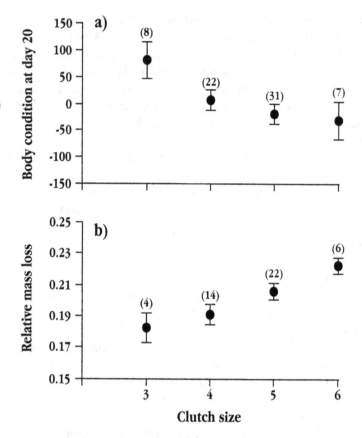

The relative importance of egg production, incubation and chick-rearing costs was studied in the common tern (*Sterna hirundo*) using an ingenious experimental design in which some nests were given an extra chick (*extra rearing costs but no extra production costs*), some were given an extra egg (*extra incubation and rearing costs*), and some birds were induced to lay an extra egg (*extra laying, incubation and rearing costs*) (Heaney & Monaghan 1995). Only pairs given a free chick reared significantly larger broods than unmanipulated controls (Fig. 2.5). There was no effect on hatchling size as a result of increased egg production, but parents bearing the full cost of the extra young showed reduced chick provisioning, resulting in reduced chick growth and survival.

Experiments have now demonstrated trade-off effects between reproductive effort and lowered survival or fecundity (reviewed in Lindén & Møller 1989). Some have also revealed the nature of underlying fitness costs. For example, in zebra finches (*Taeniopygia guttata*) one cause of this increased risk is shown to be depressed immune response, which is found to be correlated

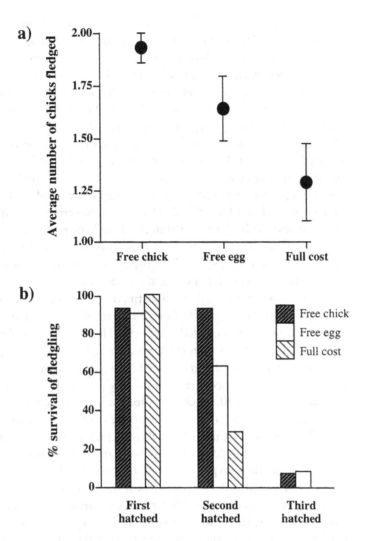

Figure 2.5
Analysis of nesting costs
in the common tern
(*Sterna hirundo*) (Heaney
& Monaghan 1995).
(a) The mean ± 1 S.E. of
chicks successfully fledged
in relation to treatment
group free chick, free egg
or full cost. (b) The
percentage of chicks
surviving to fledgling in
relation to hatching order
shown for the three
treatment groups.

with experimentally enlarged clutches (Apanius *et al.* 1996), and raising reproductive costs for blue tits (*Parus caeruleus*) was shown to reduce winter survival by producing less well-insulating feathers at the autumn moult (Nilsson & Svensson 1996). Still absent from these calculations is the possible cost of nest building, either directly in terms of construction costs or indirectly through the consequences of nest insulation on incubation costs.

2.6 The nest and clutch size

The above investigations provide evidence that the evolution of an optimum clutch size may have been influenced by a variety of factors, but what of evidence that clutch size has been influenced by the nest itself? This is an incomplete picture, but enough studies have been conducted to show that the scarcity of a nest cavity or the size of the nest itself may influence clutch size. Studer (1994), in an extensive study of forest and non-forest habitats in Brazil, found 65% of species building a cup nest, 22% building a domed nest, and 13% building a cavity nest; these experienced, respectively, 23%, 28% and 49% nesting success. Complex or closed nests in a wooded habitat in Argentina were found to have higher clutch survivorship than cup nests (Mason 1985). This is also true when comparing survivorship of clutches between closed and open nest builders in the humid tropics (Ricklefs 1969). Neotropical cavity or enclosed nesters also lay larger clutches than do open nesters (Snow 1978, Skutch 1985), a difference that also holds for European species (Saether 1985). There does, therefore, seem to be a possible relationship between clutch size and nest design.

Martin (1993b) showed that a distinction should even be made between cavity nesting species that excavate their own nests and those nesting in holes created by others. The former have significantly smaller clutches than the latter. Three hypotheses are proposed to explain this difference. The first is that the cost of excavation reduces the energy available for egg laying. However, this cannot explain the difference since, for example, clutches are still not enhanced in species in which the male carries out the excavation. The second is that it is the consequence of reduced nest predation, but observed predation levels are the opposite of that predicted. The third hypothesis is that limited availability of suitable sites for adopted cavity nesters leads them to invest more heavily in reproduction when the opportunity arises, whereas excavators can create their own nest sites. Martin (1993b) favours this explanation which, if true, demonstrates that the properties of the nest site itself can influence clutch size.

Other studies provide more direct evidence of a relationship between clutch size and nest design. Kulesza (1990) demonstrated correlation between nest design and clutch size after removing the effects of latitude and predation. For given values of nest predation and latitude, small pensile nesters lay smaller clutch sizes than open cup nesters, and open cup nesters lay smaller clutches than domed nesters. Size of the nest floor area in the cavity-nesting European collared and pied flycatchers was found to be positively correlated with clutch size by Gustaffson and Nilsson (1985), although not correlated in a study by Alatalo, Carlson and Lundberg (1988). The relationship between nest cup size and clutch size demonstrated in the barn swallow (*Hirundo rustica*) is interpreted by Møller (1982a) not as the nest determining the clutch, but as the female building a nest to accommodate the clutch she is able to lay.

The construction of extremely small nests in some tropical species, for example the grey-rumped and crested treeswifts (*Hemiprocne* spp., Hemiprocnidae) (see Fig. 6.1) or the bare-necked fruitcrow (*Gymnoderus foetidus*, Tyrannidae), where the nest can hold only one chick and the female probably stands astride the nest when incubating (Lack 1956, Béraut 1970) or cotingas such as *Iodopleura* (Sick 1979, 1993) can be interpreted as the nest determining clutch size rather than the reverse. Snow (1978) considered that such nests were to avoid detection by predators. Lill (1974) concluded that the small pensile nest of the white-bearded manakin (*Manacus manacus*) might be an anti-predator adaptation or possibly an adaptation to the scarcity of suitable building materials.

A positive relationship between nest size and predation risk was, however, demonstrated experimentally by Møller (1990a) using model blackbird nests baited with quail eggs. Slagsvold (1989a) examined the hypothesis that clutch size in some altricial birds may be limited by overcrowding during the nestling phase. Open nesting species that build nests on the ground were found to have relatively wider nest cups than off-ground-nesting species, suggesting that the design of off-ground nests limits cup diameter. Slagsvold (1989a) conducted experiments on three passerine species which supported the idea that brood size may be constrained by nest size. In the chaffinch, the number of fledged young was enhanced in a nest with a broadened rim surrounding a normal cup, and nests with an enlarged cup were even more successful. In the fieldfare (*Turdus pilaris*) an enlarged nest cup permits a larger clutch to be reared because of reduced crowding in the chick-rearing phase, but clutch size can also be related to nest material (Slagsvold 1989a). Open nests near the ground have wider cups than those above ground, and inner nest cup width is correlated with the amount of moss used in nest construction: the more moss, the narrower the nest cup. Slagsvold (1989a) suggests that there is selection for smaller nests for one or more of the following reasons: detection by predators, costs of construction, more efficient incubation. Mossy nests can overcome these problems because of the ability of the material to stretch as the chicks grow. However, moss is unsuited to exposed wet habitats because of its bad drainage properties, hence the more grassy form of the nest; but, this resists stretching, so limiting the possible size of the clutch.

The current evidence is that nests only exert a marginal influence on clutch size. This perception could very possibly change with a change in research emphasis from eggs to nests. The interest in clutch size in birds has been driven by research on life history theory; in this the role of the nest has seemed marginal. The importance of nest building costs to the total cost of reproduction is still little known (see Chapter 6), but it seems certain that a greater concentration on the properties of the nests, for example their size, shape and material composition, will uncover important relationships between the nest and key aspects of reproductive biology.

CHAPTER 3

Standardising the nest description

3.1 The nest profile survey

Nests are materials arranged in a particular way. Understanding the process of construction of a nest (Chapter 4) would therefore benefit from a detailed description of what materials appear and how they are arranged. This is equally true for speculation on functional design (Chapter 5) or on nest evolution (Chapter 9). The premise for this, however, is that nest architecture is species specific. Fortunately, evidence from a variety of animal builders gives support to this assumption.

Variations in nest architecture have been used to assist species recognition in the termite genus *Apicotermes* in which the morphology of the termites themselves shows little evidence of distinctiveness (Schmidt 1955). Taxa identified in this way have been termed ethospecies, because the clearest evidence of speciation comes from the behavioural record. Stenogastrine wasps of the genus *Eustenogaster* exhibit this pattern, with varied yet distinctive nest morphs apparently representing several species in the morphologically rather uniform genus (Hansell 1984, see Fig. 1.4). In bird literature there is a general acceptance that nests have species-specific characteristics (Collias & Collias 1984), and bird natural history books may be found in which descriptions of nests alone provide the basis for species identification (e.g. Harrison 1975). The problem with such published nest descriptions, apart from their brevity, is the lack of any systematic scheme for describing the composition and arrangement of nest materials and other key nest characters such as their dimensions or weight. The purpose of this chapter is to describe a quite detailed nest profiling scheme, which allows nest characters to be recorded simply and objectively.

Inevitably there are some difficulties in this approach which may lead to errors of categorisation or oversimplification. The most obvious of these are discussed here; nevertheless, there are two very obvious bonuses to this approach. The first is a more numerical approach to comparisons between and within species; the second is that the procedure is transparent and so capable of simple corrections where its performance is poor. The power of this approach is shown in the cladistic analysis of the evolution of nests in the social wasps (Vespidae). Wenzel (1991, 1993) applied the principle that ontogeny repeats phylogeny to the evolution of nest building behaviour, and hence species differences in nest architecture of the social wasps. The detail present in the nest architecture was sufficient to generate a phylogeny broadly corresponding to that based on the traditional characters of wasp anatomy.

The nest profile sheet described in this chapter is used to record details of 518 specimens in the museum nest survey from 44 non-passerine and 32 passerine families (Table 3.1). Data from these are presented (particularly in Chapter 5) to provide descriptive statistics that reveal underlying general relationships, for example between nest weight and the use of particular materials.

The profile for each species was obtained from a single museum specimen. It was necessary to make use of museum specimens in order to obtain a large amount of information quickly and easily, but this has certain clear disadvantages: in particular, the nest may be incomplete, it may have altered over time since collection, and by itself contains no information on within-species variation. Incomplete evidence of the nest attachment was a common deficiency, because museum nests are normally not collected attached to the nest site. This is rather a serious omission, which should be corrected where practicable in future museum reference collections. Deterioration with age appears not to be a problem; stored in a dry environment and handled little, nests remain much as they were on collection. The use of a single specimen in this survey was largely one of necessity, because nests in museum collections are frequently represented by one specimen. Nevertheless there is an advantage and a precedent in referring to a single specimen as if it were the 'type' representative for the species.

One further limitation of using museum material is that only features which are readily visible can be scored, because specimens cannot be examined destructively. For most cup nests this is not a major limitation, but for domed nests, some interior features may be missed. It should probably be said in passing that, having seen three national collections in London, Paris and Washington, and four other collections, only the impressive collections of the Western Foundation for Vertebrate Zoology in California and the National Museum of Natural History, Washington, appear to have made a systematic effort to collect nests rather than acquire them incidentally when eggs have been collected. If bird nests are to

Table 3.1. **Museum Collection Nest Survey**

Passerine family	Number of genera		Passerine family	Number of genera	
	Total	Surveyed		Total	Surveyed
Pittidae	1	1	Muscicapidae	69	20
Eurylaimidae	8	5	Sturnidae	38	10
Tyrannidae	146	49	Sittidae	2	1
Thamnophilidae	45	8	Certhiidae	22	11
Furnariidae	66	15	Paridae	7	3
Formicariidae	7	3	Aegithalidae	3	2
Rhinocryptidae	12	2	Hirundinidae	14	7
Ptilonorhynchidae	7	1	Regulidae	1	1
Meliphagidae	42	5	Pycnonotidae	21	4
Pardalotidae	16	1	Cisticolidae	14	5
Eopsaltriidae	14	2	Zosteropidae	13	1
Laniidae	3	2	Sylviidae	101	22
Vireonidae	4	3	Alaudidae	17	5
Corvidae	127	65	Nectariniidae	8	3
Bombycillidae	5	3	Passeridae	57	21
Cinclidae	1	1	Fringillidae	240	91

The nest survey covered the 32 families of passerine birds listed. The study was conducted at the level of genus and the table shows the number of genera examined in relation to the total recognised by Sibley and Monroe (1990). This amounts to a 32% sample of passerine genera. (Nests of 16% of non-passerine genera from 44 families were also examined.)

provide research material for the future, this situation needs to be changed. Great care is obviously needed to avoid collecting nests that are in use, but the nests of many species remain to be described in any detail, let alone collected.

Information was recorded under four main headings: 1) nest identification, morphometrics and type; 2) the functional zones of the nest; 3) the material composition of the nest; 4) other information. Many of the data were recorded as nominal data by ticking boxes to record the presence of features or materials. Other data were recorded as ordinal (amount of silk in the nest) or interval (dimensions and weight) data.

3.2 Nest identification, morphometrics and type

a) Identity of the nest

This section records aspects of the identification of the nest. These are in the form of scientific as well as popular names, the family to which the species belongs, both according to Sibley and Monroe

(1990) and Howard and Moore (1991). The identity of the museum and collection reference number were also recorded.

b) Nest weight and dimensions

Where possible, each nest was weighed, but this was impossible when the only nest available was attached to its original branches. Nests were weighed in the conditions where they were stored, which, for the purposes of preservation, needs to be in a dry environment. The weights are therefore likely to be less than before the nest was collected. Nevertheless, these measurements give essential indications of relative nest weight. By this method, the nest of the *Antillean crested hummingbird (*Orthorhyncus cristatus*) weighed 0.8 g, appreciably lighter than the 2.4 g bird that built it, whereas the dry weight of the nest of the *rook (*Corvus frugilegus*) was found to be 2600 g, much greater than that of the 490 g bird (Dunning 1993) that built it.

The external nest dimensions were recorded as *nest diameter* and *nest depth*. The former is the distance across the widest part of the nest. The diameter was determined by imagining the nest held loosely in a large calliper, so that sticking out branches would be excluded; even so, there is some uncertainty about precise measurement. The nest depth is the exterior top-to-bottom measurement, from the cup lip in the case of cup nests and from the top of the roof for domed nests (Fig. 3.1).

The interior measurements were *cup diameter* and *cup depth*. These are generally quite easy to measure for cup nests, where fine materials and precise workmanship minimise uncertainty. However, the point in the reproductive cycle that the nest is collected may affect the measurement, as cup dimensions are known to change with chick development (Slagsvold 1989a). For domed nests, the measurements were made at the level of the bottom of the entrance hole. Additional complications were raised by dome and tube nests, the dimensions of which were measured as in Fig. 3.1. It is therefore clear that dimensions of the same nest design (cup, dome, dome and tube) can be compared between species; comparisons between designs are not meaningful.

A small number of nests had substantial amounts of material extending above the roof or below the bottom of the nest (Fig. 3.1). These are treated as separate nest features, termed *head* and *tail*, the lengths of which were measured. The possible function of these features is discussed in Section 5.2d. Where nests were collected attached to a branch, the width of the support at the point where the weight of the nest is borne by a single branch was measured as

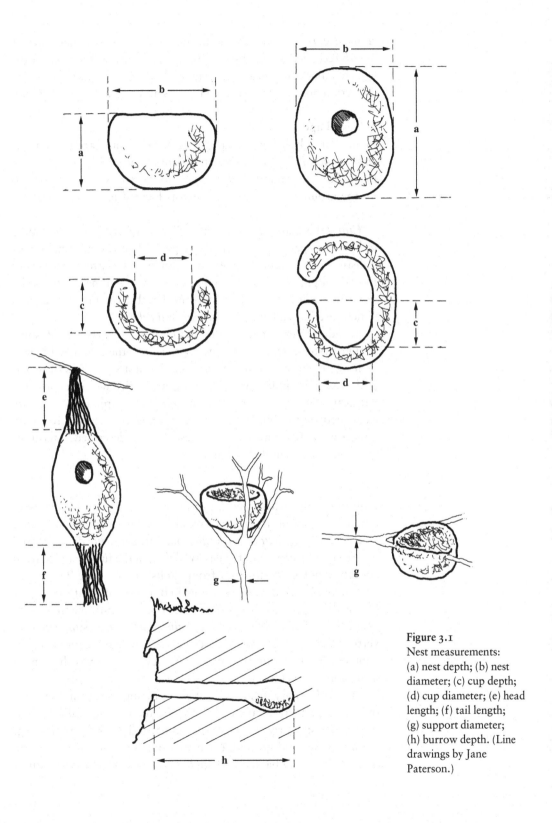

Figure 3.1
Nest measurements:
(a) nest depth; (b) nest
diameter; (c) cup depth;
(d) cup diameter; (e) head
length; (f) tail length;
(g) support diameter;
(h) burrow depth. (Line
drawings by Jane
Paterson.)

support diameter. Finally, nests classified as burrows had *burrow length* recorded as the distance from the mouth to the back of the egg chamber. This measurement could only be obtained from field collection notes in the museum records and so was relatively rare.

c) Nest shape

Eight mutually exclusive categories of nest shape are recognised: *cup, dome, dome and tube,* and *plate* all of which are generally located above ground; *bed, scrape* and *mound*, which are located on the ground, and *burrow*, which is located in the ground (Fig. 3.2).

Cup is the shape of a nest with a distinct or prominent concavity to hold the eggs, whereas an above-ground nest with a shallow or indistinct cup is termed *plate*. An example of the latter is the nest of the *woodpigeon (*Columba palumbus*). A roofed nest is termed a *dome* unless it has an additional antechamber or entrance tube, in which case it is categorised as *dome and tube*.

Bed describes a flat or shallow cupped nest typical of some ground nesters; *scrape* a shallow depression in the ground with little or no gathered material; and *mound* the nest structures of megapodes (Megapodiidae), where the eggs are buried within a constructed heap of material. The category *burrow* is applied to a cavity excavated by the bird either in the ground or tree; where cavity nesters use ready-made sites, the nest shape is described in terms of the structure assembled within it.

d) Nest site

Eight mutually exclusive categories of nest site are recognised: *tree/bush, grass/reeds, ground, tree hole/cavity, ground hole/cavity, wall, ledge* and *water*. An additional category, *no data*, is used where no evidence of nest site is available, in the first instance from collection notes or ultimately from published literature (Fig. 3.3).

Tree/bush includes any nest sited in branches of whatever size or height above ground. *Grass/reeds* encompasses nests like that of the *Eurasian reed warbler (*Acrocephalus scirpaceus*), supported on vertical stems; while *ground* describes nests which are physically in contact with the ground or raised only a few centimetres above it on a cushion of vegetation.

Tree hole/cavity includes cavities occurring in rotting branches or trunks, either excavated by the nester or adopted. *Ground hole/cavity* similarly embraces sites in natural ground cavities (e.g. among rocks) and ground, excavated burrows used either by the diggers or second hand. *Wall* applies to a nest attached to natural

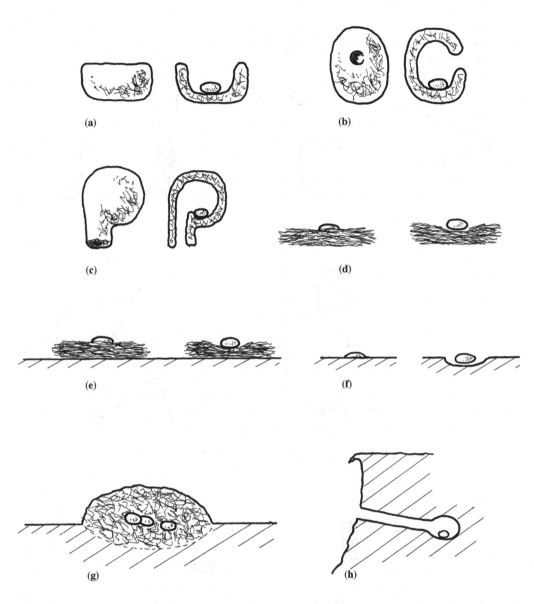

cliffs or the walls of buildings where it receives little or no support from below. *Ledge,* however, describes sites on cliffs or buildings where the nest does receive substantial support below it (Fig. 3.3). Finally, *water* is confined to nests made of materials which allow them to float.

e) Nest attachment

The definition of an attachment is that it is a device for holding a nest in position. Each nest was assigned to a single attachment

Figure 3.2
Nest shape was recorded as one of eight possible categories: (a) cup; (b) dome; (c) dome and tube; (d) plate; (e) bed; (f) scrape; (g) mound; (h) burrow. (Line drawings by Jane Paterson.)

Figure 3.3
Nest site was recorded as one of eight possible categories: (a) tree/bush; (b) grass/reeds; (c) ground; (d) tree hole/cavity; (e) ground hole/cavity; (f) wall; (g) ledge; (h) water. (Line drawings by Jane Paterson.)

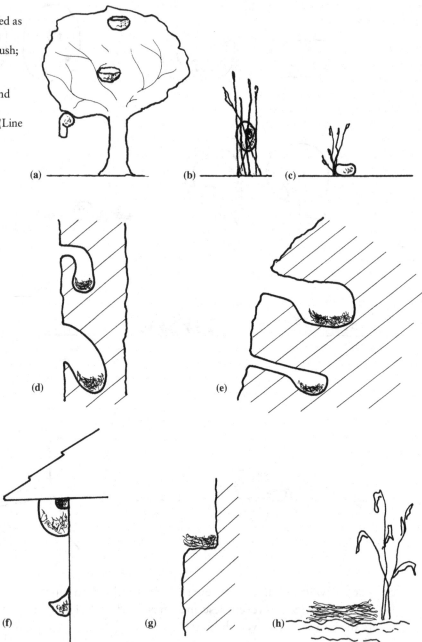

category. Where present, the attachment materials were also re-corded. However, according to the scoring rules, not all nests have attachment materials. For example, a twig nest like that of the *rook, located between forked branches, may be locked in position by virtue of nest twigs projecting between the supporting branches, yet not have any special attachment. Grassy or mossy nests may be similarly secure in forked branches without any attachment device. Yet, if the materials are built around or over the tops of branches or other features of the nest site, they are recorded as components of the nest attachment. Ten mutually exclusive categories cover the range of possible nest attachments: *top, top lip, top side, bottom side, wall, bottom multiple (branched), bottom multiple (vertical), bottom, leaf purse, on ground* (Fig. 3.4). In addition, the category *no data* is needed in the absence of any information on the nature of the attachment of the nest to its site.

On ground includes all nests placed on the ground, raised on a small cushion of vegetation or placed on a rock ledge. These need no special feature to hold them in position and, by definition, have no attachment materials. The remaining nine categories describe methods for nests sited above ground. *Bottom multiple (branched)* describes nests in trees or bushes which are supported from below by contact with two or more branches. In such a site a nest may be secured simply by its weight, which lodges it in the angle of the branches; alternatively, there may be attachment materials to pre-vent it being dislodged. *Bottom multiple (vertical)* describes those nests attached to several vertical stems of reeds. To stop such a nest slipping down, the nest wall is built around the stems or bound to them with attachments projecting from the wall (Fig. 3.5). The attachment of a nest supported from below by a single horizontal branch is described as *bottom*. Nests of this type generally rest on a branch of diameter similar to their own, but which may be substan-tially less, in which case some special attachment materials are required to prevent them toppling from their position (Fig. 3.6). Nests attached to a single vertical branch or one steeper than a 45° angle were categorised as *bottom side*, while nests attached to the side of a drooping leaf were designated *top side* because the support of the nest derives from above rather than below it. For both of these, some attachment material is essential.

Two distinct categories of suspended nests are recognised: *top lip* and *top*. The former describes open, purse-shaped nests supported by branches attached at the nest lip. This may be along two sides in the angle of forked branches (also referred to as *hammock*, see Section 5.3c), or at a single point on the nest lip. *Top* attachment

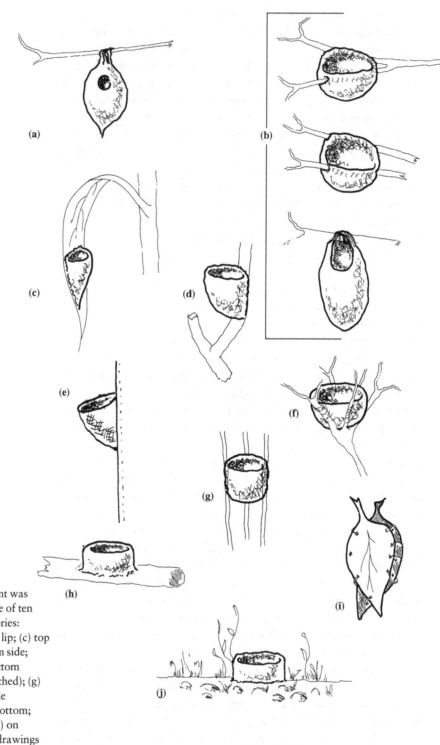

Figure 3.4
Nest attachment was recorded as one of ten possible categories: (a) top; (b) top lip; (c) top side; (d) bottom side; (e) wall; (f) bottom multiple (branched); (g) bottom multiple (vertical); (h) bottom; (i) leaf purse; (j) on ground. (Line drawings by Jane Paterson.)

Figure 3.5
Bottom multiple (vertical) attachment: detail of the nest of the *scarlet-headed blackbird (*Amblyramphus holosericus*, Fringillidae) showing how the nest is attached to a vertical stem. (Photograph by M. Hansell.)

refers specifically to domed nests, with or without an entrance tube. Both require specialised attachment materials. Finally, *wall* describes nests attached to a more or less vertical rocky surface or wall, requiring some kind of adhesive as an attachment material, and *leaf purse* describes the living attachment of the specialised nests whose structural walls are the living leaves of a bush held together by stitches or similar fastenings.

3.3 The four nest zones

The composition of the nest is described for four functionally distinct areas of the nest, which are generally reasonably easy to recognise, although all four may not be present in all nests. They are, from exterior to interior: *attachment, outer (decorative) layer, structural layer* and *lining*.

The most important of these is generally the *structural layer*. The critical feature of it is that it gives integrity to the nest shape, preventing it distorting or falling apart. Any component of the nest considered to be performing this role is treated as a component of the structural layer. In the majority of nests examined, the

Figure 3.6
Nest of *Costa's
hummingbird (*Calypte
costae*) showing bottom
attachment of a
high-walled nest. To
prevent toppling of the
nest from the site, silken
'guys' are stretched from
the nest wall to the
branch. (Adapted from
Rowley 1966.)

structural materials appeared as a single layer of more or less
uniform consistency. However, nests may have more than one
structural layer, each of a different composition. Kulczycki (1973)
describes the nest of the common raven (*Corvus corax*), for
example, as composed of three structural and two lining layers.
Some nests in the survey were found to have more than one structural layer or one composed of more than one element, each of
different construction (e.g. a base platform and a ring-like wall). In
such cases all the materials were recorded under *structural layer*,
but notes were added under *Special features and comments* describing the arrangement of these component parts.

Where arthropod silk is a structural component it is scored on a
three-point, ordinal scale, where 1 represents a detectable but small
amount contributing slightly to the strength of the nest, and 3
describes abundant silk upon which the nest structure is entirely
dependent. On this scale, a score of 0 can be added to denote the
absence of any structural silk. In some cases the structural layer is
treated as *absent*, for example a ground or cavity nest may have no
materials other than those used for other purposes, for example
insulation, and is therefore classified as *lining*. The four nest areas
are mutually exclusive, so a material classified as structural cannot
be in the *lining layer* even if it is in contact with the eggs or chicks.
Only a distinct layer inside the structural layer but with no significant structural role can qualify as *lining*. By this definition, a lining
layer is often absent.

Attachment records the presence of any materials used to secure
the nest in position. Where no attachment materials are used, this is
recorded as attachment *absent*. A nest supported from below may
or may not have attachment materials, but a nest not supported
from below much have at least one. Where silk is used as an

attachment material its importance is recorded on a three-point scale as for the structural layer.

The fourth recognised nest zone, *outer (decorative) layer*, describes material that is placed on the outside of the structured layer, altering the nest appearance without contributing to its strength or site attachment. In the nests of many species no such materials are found and the layer is described as *absent*. When present, the materials typically include lichens and little else. The function of such materials and whether their purpose is only or partly to alter the visual effect of the nest are discussed in Section 5.2.

3.4 The materials

The nest of the Florida grasshopper sparrow (*Ammodramus savannarum floridanus*, Fringillidae) is described by Delaney and Linda (1998) as containing 12 different plant materials, ten of them identified to species. The paper goes on to characterise the nest as mainly composed of 'narrow leaved grasses and grass-like monocots', which would be regarded as one material in more general nest descriptions. The scoring system developed for the nest survey produces categories which are based on the properties of the materials as well as their origin. It recognises that, for a builder, the distinction between, for example, grass leaf and grass stem may well be more important than that between one grass species and another (see Section 5.4f). However, there is a problem of how many categories to create. Oniki (1975), in a study on slaty antshrikes (*Thamnophilus punctatus*), scored four types of rootlets in a scheme that included only 13 categories of material. These differences may be of significance in the structure of the nest. The species of moss and lichen in the nests of *long-tailed tits (*Aegithalos caudatus*) are apparently chosen for their characteristic morphology and hence abilities to perform a specific function (Sections 4.7c and 5.4c). In the nest survey there was only one category of rootlets, taking the view that additional categories might be meaningful but only appropriate to a more detailed investigation. The result is a scheme that is intended to be practicable and to identify categories from the builders point of view.

Thirty-three specific categories of building material were recognised and recorded as nominal data (present or absent: two mineral, seven animal, and 24 of plants, lichen and fungi). The number chosen to some extent reflects the nature of this particular survey, yet it also maintains the intention of capturing a nest's essential character. Four nest layers of 33 materials gives 132 boxes in which materials can potentially be recorded. In practice, nests proved to be composed of few materials, with one or more zones frequently unrepresented in the nest at all (see Fig. 5.1).

Each nest had to be examined non-destructively, aided only by a powerful hand-held magnifying glass. In some cases, very small samples of material were removed for subsequent, more detailed examination, but for the most part categorisation was immediate. To be included, a material was required to be making a significant contribution to the nest. This has the effect of excluding materials that appear as just a few pieces, and it reduces the time needed to examine each nest yet ensures that all materials that should be, are included.

The greatest variety of materials, and indeed the most important materials in bird nests overall, are of plant origin; of lesser importance and variety are materials of animal origin, with inorganic materials featuring in only a minority of nests. This trend is reflected in the scoring systems in the proportions of the 33 categories that are devoted to each. One further category, *others*, was used to record the presence of any material falling outside this scheme.

a) Inorganic materials

The two categories are *stones*, a rare and unimportant material, and *mud*, a vital material in the nests of about 5% of bird species (Rowley 1970).

b) Animal materials

For some species, building materials are products of their own metabolism. This is illustrated by two materials, *salivary mucus*, as used by species of swifts and swiftlets (Apodidae), and *feathers*, like the down feathers of the female *common eider duck (*Somateria mollissima*), which she removes from her own breast in large quantities to insulate the eggs. A number of smaller passerines in temperate areas also make use of feathers, but collect those lost through moulting by other birds or dispersed at bird kills. The *fur* or *hair* of mammals, collected in a similar way, forms another animal material used in nests, as is, more unusually, cast *snake skin* (Section 5.2c).

The three remaining classes of animal material are of arthropod silk. These had to be identified from minute fragments removed from nests for later microscopic examination. This was to distinguish them unambiguously from plant down, mammal wool or synthetic fibres, which they may superficially resemble; this did, however, also allow different sorts of arthropod silk to be recognised.

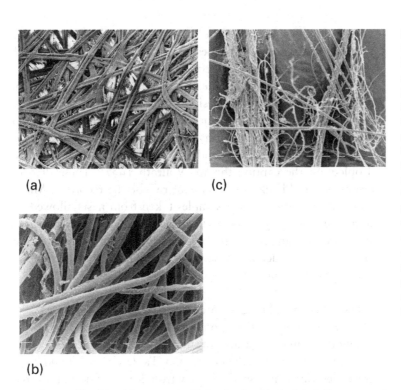

(a)

(c)

(b)

Figure 3.7
Characteristic differences
between types of
arthropod silk.
(a) Lepidopteran silk
showing diagnostic
double-stranded
appearance, taken from
the nest of the *maroon
oriole (*Oriolus traillii*).
(b) Spider cocoon silk
with typical single thread
type, taken from the nest
of the *Guam flycatcher
(*Myiagra freycineti*).
(c) Spider web or retreat
silk, characterised by
multiple thread diameters
and tangled appearance;
sample taken from the
*hair-crested drongo
(*Dicrurus hottentotus*).
(Scale: separation
between two white bars is
10 μm.) (Photographs by
Margaret Mullin.)

Caterpillar (Lepidopteran) *silk* is obtainable by birds from pupal cocoons or from bivouacs of communal-living species. It can be distinguished because it is produced by the extrusion of silk through a pair of labial spinnerets to produce uniformly paired strands of silk (Fig. 3.7). Spiders produce silk generally through three pairs of abdominal spinnerets, with which are associated a number of different specialised glands (Vollrath 1992). Spiders use their silk to make capture devices, retreats, and cocoons to cover their eggs. The composition of silk in cocoons is generally simple and its arrangement compact rather than elongate. It was often possible confidently to classify material as *spider cocoon silk* (Fig. 3.7), a material that proves to be quite important in bird nests. The remaining silk samples generally had strands of varied diameter, confirming their spider origin. All these samples were designated simply *spider silk,* the origin of which might be capture webs of whatever design or retreats. This gave three categories of silk for general analysis.

There are two types of capture silk used in spider webs: those made by cribellate and by ecribellate spiders. Cribellate spiders produce a mass of very fine threads (about 0.015 μm in diameter) combed onto a much thicker core thread. These very fine threads are extruded through a cribellum, a structure found on spiders of

certain families, notably Uloboridae, Amaurobiidae and Dic-
tynidae (Eberhard 1988). These capture threads can be readily
recognised in electron microscope examination, and their capture
principle is entanglement of the spines and hairs on insects or by an
unknown mechanism of adhesion to the shiny surfaces of insects
(Opell 1994). The capture principle of the large orb web spiders of
the family Araneidae, and illustrated by the New Guinea ladder-
web spider (see Fig. 1.1), depends upon adhesion from viscid, sticky
droplets on the capture thread (Vollrath 1992). These droplets
disappear in old silk and so cannot reliably be recorded. Microscopic examination of silk samples taken from nests allowed the
spider silk category to be divided into three: *spider web*, *spider
cribellate web*, and *spider unclassified silk*. This gave five categories
of silk in a more detailed analysis of the frequency of silk types in
selected bird families (see Table 5.4).

c) Plant, lichen and fungal materials

The 24 materials in this group are distinguished not simply on
taxonomic criteria, but also on the part of the plant from which
they come, and even on their condition (Table 3.2). So, *broad leaves*
are as described and derived largely from leaves of deciduous trees
and shrubs, whereas *broad leaf skeletons* are the lacy remains of
leaves retrieved from the leaf litter, and *leaf petioles* are the shafts of
the pinnate leaves of deciduous trees gathered after they have fallen
and the pinnae have been cast. *Pine needles,* a distinctive material,
is self-explanatory.

Stem material is distinguished as *fine woody stems* regardless of
the species from which the stems originated provided they are less
than about 2 mm in diameter and flexible. *Sticks* refers to stems of a
greater diameter and characterised by their stiffness. *Vine tendrils*
are of variable stem diameter, but diagnosed by their corkscrew
form. *Rootlets*, generally similar in their diameter and flexible
quality to fine woody stems, are identifiable by their rather wavy
profile and flimsy, short side branches.

Two categories, *bark* and *plant down*, have the advantage of
being easy to recognise, but are unsatisfactory in certain other
respects. *Bark* embraces material of quite varied character, ranging
from pieces that are stiff and brittle to those that are soft and
flexible, giving an impression of uniformity in a category that is
varied in its qualities. *Plant down*, on the other hand, has a rather
uniform quality of being soft and fluffy, but combines materials of
broadly two different origins: plant hairs that occur as a pappus to
aid seed dispersal, and hairs that occur on stems and leaves to

Table 3.2. The 24 recognised categories of plant, lichen and fungal material recorded as present or absent from each of the four nest zones: outer (decorative) layer, attachment, structural layer and lining layer

1. Broad leaves	13. Grass stems
2. Broad leaf skeletons	14. Grass heads
3. Leaf petioles	15. Palm fronds
4. Pine needles	16. Palm fibres
5. Fine woody stems	17. Green plant material
6. Sticks	18. Flower heads
7. Vine tendrils	19. Horsehair fungus
8. Rootlets	20. Fern stem scales
9. Bark	21. Moss
10. Plant down	22. Lichens
11. Grass leaves (narrow)	23. Rushes
12. Grass leaves (broad)	24. Seaweed

protect the plant against desiccation or herbivorous insects. With reasonable botanical knowledge and microscopic examination of samples, this category could be made into two.

A problem was also caused by the use of grass leaves in nests because the large family Gramineae, the grasses, range from fine meadow grasses to 20 m tall forest bamboos (Bambuseae). The leaves of grasses may be used by birds either entire or as strips. This was resolved by creating two grass leaf categories on the basis of the width of the piece of leaf employed. *Grass leaf (narrow)* describes a leaf or leaf strip less than 5 mm width, while *grass leaf (broad)* applies to all leaves or strips of greater width. In practice, leaves of the related family Cyperaceae (sedges) may have been included in one or other of these. Both are Monocotyledoneae and so characterised by long parallel veins in their leaves.

Grass is a widely used material, reflecting its extensive availability in a great range of habitats, but birds may distinguish different parts of the plant, selecting the part with properties suited to a particular function. Consequently, two further grass categories were required, *grass stems* and *grass heads*. The latter was generally of the head attached to a long stem, but included heads that are compact or loose and so somewhat varied in character. Creating four grass categories seems like making distinctions without significance, but this is not the case; for example, a grass leaf has the structure of a flexible ribbon, whereas a grass stem is designed as a stiff, hollow beam. This demands differences in building technique and suits them to rather different structures (Section 5.4f).

An attraction of the grass leaves as building materials is that they can be used as long narrow strands, either as whole leaves or in the form of strips torn from broader leaves. In the tropics and sub-tropics, another large family of monocots provides similar opportunities, the Palmae (palms). *Palm frond* describes pieces that can be as broad as a whole leaf or narrow strips torn from a frond, whereas *palm fibres* describes strands like fine wires that may, for example, be collected from around the trunks of some palms or possibly stripped from partially rotted fronds.

A category *green plant material* is included to identify fresh plant materials that might contain secondary compounds with insecticidal or antibacterial properties (Section 6.9); however, in the nest survey it proved impossible to use this reliably on museum material unless supported by published confirmation. Consequently, some materials collected while green may be underestimated. This might be true of some records of *flower heads,* a category which describes racemes of dicot herbs in nests for example of some shrikes, Laniidae (*lesser grey shrike, *Lanius minor*).

Another material that takes the form of fine, long strands and offers building opportunities similar to those of grass or palm strands is so-called *horsehair fungus* (*Marasmius*) (Basidiomycetes). This may be found attached on the surface of leaves, where it creates a tangle of blackish, wiry rhizomorphs (bundles of hyphae), in tropical forest (Hedger 1990) and also in some temperate forests (McFarland & Rimmer 1996). Ferns (Pterophyta) provide another specialist material, *fern stem scales.* These are the scales that envelop the frond before it uncoils and then remain attached to its stem; they have a very characteristic papery quality and elongate, pointed shape (Fig. 3.8). More predictable, although again somewhat variable in structural detail, are *moss* and *lichen.*

The remaining two plant categories cover the most characteristic materials of aquatic origin. *Rushes* was a category without a precise taxonomic identity and used to describe stems and leaves of emergent aquatic vegetation of monocot families like the Typhaceae or Iridaceae. Some of this vegetation, by virtue of the air spaces inside it, floats and so can be used to construct floating nests. *Seaweed* covers all marine algae, and is a common material for cliff and shore-nesting sea-birds.

d) Others
Having this additional category allows attention to be drawn to the presence of one or more materials not included in the list described. This could be an additional type of animal or plant material: for

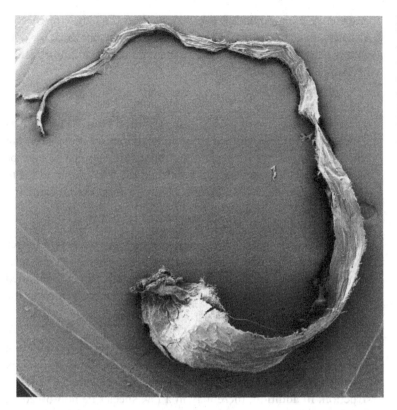

Figure 3.8
The stems of ferns and tree ferns bear papery, triangular and elongated scales used in nest construction by some species, for example the hummingbird *Eulampis jugularis*. (Photograph by Margaret Mullin.)

example cow dung in the lining of the nest of the *spotted morning-thrush (*Cichladusa guttata*, Muscicapidae), or a human synthetic material such as paper as a decorative material on the outside of nests of the *rufous-browed peppershrike (*Cyclarhis gujanensis*, Vireonidae). The actual nature of these materials can then be described in the *Special features and comments* section.

3.5 Additional information

a) Building techniques

The building techniques revealed in the nest structure are recorded by the presence of six categories: *sculpting, piling up, moulding, sticking together, interlocking* and *weaving*. These categories are based on those of Hansell (1984), but modified to a new understanding of construction principles which is described in Section 4.2.

Recording these categories proved not entirely satisfactory in the museum survey because the categories are not mutually exclusive, and to confirm all techniques employed in one nest would require destructive examination. Nevertheless, recording the presence or absence of each of them provides an objective framework for data collection which could readily be achieved in a more detailed study of a smaller number of species.

b) Adults and young

There are a number of pieces of information relating to the nest which cannot be determined from an examination of the nest, but which are important for understanding its biology. These may be found in scattered or compiled published sources and so added to the nest profile sheet if available. The following were included in this survey.

Weight of the adult bird. This was the single most important source of additional information in this survey because it allows the relationship between bird weight and nest weight to be examined (Chapter 5). In this survey, the bird masses were taken from the compilation of Dunning (1993).

Who builds. One or both sexes may be involved in nest building and when it is both there may be division of labour, with one sex largely being concerned with collection and the other with construction. In addition, there may in a few species be non-breeding helpers participating in nest construction. Scattered information on this is available in the literature on a species-by-species basis, and a broad family-by-family compilation is provided by Collias and Collias (1984). The data are, in principle, very interesting because who builds the nest raises important questions about the role of the nest in sexual selection. This is given some consideration in Chapter 8.

The clutch. The categories included here were *clutch size, egg size* and *egg weight* (all available either as scattered publications or in museum collection records). None of these was available in a sufficiently comprehensive form to allow further analysis; but, in principle, it is important to know the volume of nest cup occupied by the eggs and the total weight of clutch plus incubating adult, for example in relation to nest suspension.

c) Biological associations

Eight possible categories of association could be recorded as present or absent: nest associated with *nest of conspecifics* (i.e. colonial), *nests of other bird species* (notably near to nests of raptors), *nests of bees*, of *wasps*, of *ants*, and of *termites*. These are considered as aspects of nest site selection in Sections 7.6–7.12.

The two further categories of nest association are association with *humans* and *other*. The human category proved unusable because, although association with human habitation is reported (Collias & Collias 1984), there is no evidence that birds nest in such sites to be near humans or obtain any benefit from it. *Other* simply allows additional associations to be recorded should they appear, but none was found in this survey.

d) Special features and comments

One box was used to record simply the presence or absence of any further notes or comments. Where this was *yes*, a short note of text was added in an additional box. Examples include recording that two rather than one structural layer could be distinguished, or that the structural layer was built in two parts, a shallow bowl of one material bound round with a high-walled ring of another (e.g. *greater Antillean bullfinch, Loxigilla violacea*, Fringillidae). These provide additional detail of structures whose composition is already recorded. *Top-lip suspension but no silk* (*Attila spadiceus*, Tyrannidae) and *hanging mud nest* (*Lochmias nematura*, Furnariidae) simply draw attention to an unusual combination although already scored in the appropriate boxes. *Nest built on a previous nest*, a behaviour characteristic of some hummingbirds (e.g. the *long-tailed hermit, Phaethornis superciliosus*), adds a character not otherwise considered in the nest profile.

e) Sketches and photographs

At the time of the examination of nests in this survey, pencil sketches were also made on the check sheet, with some brief explanatory notes. The nest was also photographed at least once. More complicated or unusual nests were sketched in more detail, to which might be added photographs of more than one aspect or perhaps a close-up.

CHAPTER 4

Construction

4.1 Introduction

Is an organism that spends only five to ten days a year exhibiting a certain type of behaviour likely to possess anatomy specialised to perform it? The key feature that has allowed birds to be such notable builders is a bodkin beak on a mobile neck close to a pair of sharp eyes. Nevertheless, evidence that the morphology of the beak or any other part of the bird is determined to any degree by building technique is weak or absent. The partial fusion of the front three toes of kingfishers (Alcedinidae) may assist with burrow excavation (Sick 1993), but such claims of morphological adaptations for nest building are hard to find.

Feeding is, day to day and year round, what beaks are used for, so this is likely to be the dominant influence on beak shape. Evidence to support this, although not always compiled in a systematic way, is very generally evident. Such a relationship has been deduced, for example, from the markedly different beak shapes of the Hawaiian honeycreepers (Drepanidini, Fringillidae) accompanying their radiation into different feeding specialisations (Buhler 1981, Fig. 4.1). The nests of the honeycreepers, however, are unremarkable cup-shaped structures, lacking in diversity. The *sturdy-billed palila (*Loxioides bailleui*) and *Laysan finch (*Telespiza cantans*), birds of around 35 g (Dunning 1993), build with fine woody stems, rootlets and grasses, whereas the finer-beaked species, the *apapane (*Himatione sanguinea*) and iiwi (*Vestiaria coccinea*), birds of 25 g, build with almost identical materials (Carlquist 1980).

So, if it is generally true that beaks are modified as feeding instruments, this could act as a constraint upon or as facilitation for certain methods of nest construction. Woodpeckers (Picidae), for example, could have adopted the excavated nesting habit as they became morphologically specialised for digging insect prey out of dead wood with their sharp beaks. The survey of nest building techniques in this chapter, however, supports the view that there is little relationship at all between beak morphology and nest building technique. A detailed research study might show otherwise; indeed,

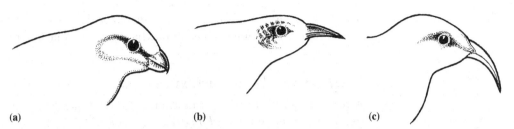

(a)   (b)   (c)

it would be interesting to look for small anatomical differences between the sexes in species in which only one is the builder, or even to look for seasonal morphological changes that could be attributed to the demands of nest building.

In the absence of such studies, the literature provides only fragmentary evidence and speculation. For example, the interpretation of the special shape of the beak of the hamerkop (*Scopus umbretta*) is either 'ideal' for the manipulation of the large twigs of which its nest is made (Liversidge 1963) or, alternatively, specialised for feeding on tadpoles (Siegfried 1975). More general claims have also been made about the relationship between nest structure and body morphology that deserve further investigation, for example the suggestions that ducks and relatives (Anatidae) are poorly adapted for picking up objects in the beak, which could have limited their scope for nest building (Kear 1970). To assess such a claim needs a detailed examination of skull anatomy, which in birds is substantially more complicated than in mammals. The latter have only one articulating joint, whereas birds have up to 13 (Buhler 1981). The number is reduced in passeriformes; nevertheless, the plasticity of their beaks is demonstrated by the evolution of a large range of forms in Hawaiian honeycreepers in as little as 10 million years (Fig. 4.1).

If, however, it is true that beak and other aspects of bird anatomy are unspecialised for nest building, this could be compensated for by the presence of specialised behaviour. Here, detailed descriptive studies, let alone experimental manipulation, are surprisingly scarce. Consequently, there is some uncertainty as to whether construction behaviour is difficult or simple, and the extent to which building skills develop with experience. This chapter provides an assessment of the complexity of building behaviour and the extent to which learning may be involved. The various construction methods used by birds are categorised and then the principle by which the nest materials are held together is examined, since some materials hold together by properties inherent in them while others require special construction behaviour. The chapter ends by comparing the complexity of nest construction behaviour with that of

Figure 4.1
The beaks of the Hawaiian honeycreepers (Drepanidini) are interpreted as adaptations to different feeding specialisations and not in relation to their nest construction behaviour. (a) The palila: short massive beak; diet of tough *Sophora* and *Myoporum* seeds. (b) The *apapane: slender, pointed beak; diet of nectar and caterpillars. (c) The iiwi: long, curved beak; diet of nectar and caterpillars.

bird tool use and tool manufacture, in particular to determine whether any special importance should be attached to the latter.

4.2 Types of construction method

A fundamental distinction in nest building techniques is that between *sculpting* and *assembling* (Hansell 1984). In the former, a shaped surface or cavity is created by the removal of material. This may be as little as a shallow scrape on the surface of the ground or a long burrow dug in a bank or cavity excavated in a tree. The latter, *assembling*, describes a variety of distinct building techniques in which materials are collected and joined together to create a receptacle for the eggs. Assembly techniques, which are shown by the great majority of nest building birds, can be divided into: *piling up, moulding, sticking together, interlocking, sewing* and *weaving* (Hansell 1984). The purpose of all these techniques is essentially twofold: to ensure that the nest stays attached to the nest site and that the components of which it is made do not fall apart.

This can be most readily understood with *piling up*, where the materials are simply laid on top of one another. Suitable materials are varied using this technique. On the ground, leaves or stones may be adequate, but in trees a typical material is sticks, which can form beams or struts to bridge gaps between branches. *Moulding* in this context is the shaping of a wet material which, on drying, hardens to the desired shape. For attachment of the nest, the moulded material acts as an adhesive securing the nest to the substrate. The materials used in this category are mud and, in a few species, orally secreted mucus. *Sticking together* describes nest building materials held together by an adhesive or glue. An adhesive is by definition applied as a liquid. Mucus, mud and faeces can all act as adhesives, together with a variety of different materials normally of plant origin.

Interlocking covers a group of rather varied techniques and materials which have the common feature that they do not depend on an adhesive because the materials, once brought in contact, stay together. Instead, the materials become ensnared in one another by virtue of simple construction movements and of the properties of the materials themselves. Three particular interlocking devices are identified: *linked coils, stitches* or *pop-rivets*, and *velcro*. A fairly small number of birds interlock coiled vine tendrils to form a nest fabric. An even more select group hold a bundle of fibres in the beak which are driven through a leaf; the tuft or knob emerging on the other side may then be driven through another leaf, linking them in a stitch, or be left as a pop-rivet, fastening the fibres to the leaf surface. A much larger number of birds use arthropod silk as a looped material into which the projecting surfaces of plant materials become hooked, so building up a *velcro* fabric.

Weaving is, somewhat arbitrarily, treated as a separate technique from interlocking. The weaving found in bird nests falls short of the regular alternation of threads of weft over and under parallel threads of warp, as in human woven textiles. What characterises bird weaving is that the vegetation strands of which the nest is composed need to be carefully intertwined, even to the extent of using loops, hitches and knots, because they do not readily interlock or bind together. The construction technique, therefore, by the nature of the material, imposes more demands upon the behaviour of the builder.

The details of behaviour recorded for each of these construction techniques can now be examined under these headings: *sculpting, moulding, piling up, sticking together, interlocking* and *weaving.*

4.3 Sculpting

Sculptors, whether in trees or in the ground, illustrate the power of a bird's beak to excavate a cavity large enough for the rearing of the young and, in the subterranean species, digging a long burrow to give them additional security. The pre-eminent examples of the sculpting technique are the woodpeckers (Picidae), a family of just over 200 species worldwide (Sibley & Monroe 1990) whose typical nest is a cavity carved into a dead tree limb with a narrow entrance for limiting access, e.g. the *downy woodpecker (*Picoides pubescens*), or, in the case of the red-cockaded woodpecker (*P. borealis*), in living pines softened by fungal infection (Ligon 1970). The method of construction in these species is to strike the wood with the sharp chisel of a beak, while gripping the trunk with powerful claws. Since this is also the method of dislodging dead wood to reveal insect prey, there is probably little about the nest construction behaviour that is unique to it, other than the sustained shaping of the hole to create a nest cavity.

For the kingfishers (Alcedinidae), however, that use rather similar chiselling actions with large pointed beaks to excavate burrows in river banks or termite nests (Hindwood 1959), the diet is vertebrate prey such as fish or lizards, consequently nest digging actions may be quite different from feeding ones. The rufous-tailed jacamar (*Galbula ruficauda*, Galbulidae) is another species that excavates a burrow in an earthen bank; it feeds on flying insects and has a long, finely pointed beak, which Skutch (1983) marvels could be suited to either feeding or digging. Certainly, looking at the beak shapes of burrowers and cavity excavators from several different families, there is little similarity in the form of their beaks; however, cavities may be excavated with the beak using other techniques such as gouging or biting. These are more suited to softer materials because the beak is not struck but pushed into the substrate. Among tree cavity nesters, the gouging technique is known for a number of species of trogon (Trogonidae), for example the *collared trogon (*Trogon collaris*), baird's trogon (*Trogon baidrii*), and the resplendent quetzal (*Pharomachrus mocinno*) (Fig. 4.2, Skutch 1983). The nest excavation technique shown by parrots (Psittacidae), whose beaks are powerful to cope with feeding on tough fruits, is biting. Hardy (1963) noted that orange-fronted parakeets (*Aratinga canicularis*) showed only discrete body and head movements while biting into the friable surface of a tree termite (*Nasutitermes*) nest to create a nest cavity.

The feet of birds have not been recorded as actual digging instruments but, for example in the kingfishers, motmots and jacamars (Skutch 1976) and the Manx shearwater (*Puffinus puffinus*) (Lockley 1942), the feet are used to kick out of the burrow material dislodged by the beak. In many of these cavity excavators the beak may also be used to collect material, because, in the case of species burrowing in the ground in particular, the floor of the nest chamber and in some cases the whole cavity are lined with vegetation.

4.4 Moulding

The majority of birds that use this technique build with mud, a material estimated by Rowley (1970) to be used by no more than 5% of the world's bird species. Only a small proportion of these build their nests largely or wholly of mud, and so can truly be said to mould the plastic material to the required shape. The only other bird nest material to be moulded is salivary mucus, secreted by the builders themselves. This mucus, unmixed with other materials, is used to construct the nest of the edible-nest swiftlet (*Collocalia fuciphaga*), these so-called *white* nests being more prized for cooking than those of the black-nest swiftlet (*Collocalia maxima*) in which the saliva is mixed with feathers (Medway 1962). There is no detailed description of the nest construction of either species.

Swallows and martins (Hirundinidae) build up the nest shape by the placement of large numbers of mud pellets to the growing nest

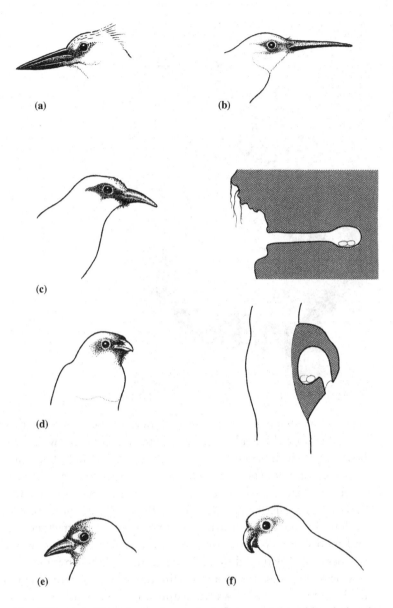

Figure 4.2
Beaks for excavating a nest burrow or cavity; beak shape is a generally poor guide to the type of nest. (a) Ringed kingfisher (*Megaceryle (alcyon) torquata*);
(b) rufous-tailed jacamar (*Galbula ruficauda*);
(c) rufous-capped motmot (*Bathyphthengus ruficapillus*) and burrow;
(d) surucua trogon (*Trogon surrucura*) and nest cavity;
(e) rusty-breasted nunlet (*Nonnula rubecula*);
(f) peach-fronted parakeet (*Aratinga (aurea) aurea*).
(b) and (c) typically dig burrows in banks; (d) and (f) excavate cavities in the nests of arboreal termites; (a) and (e) nest in cavities dug in the mounds of terrestrial termites.

rim (Fig. 4.3). This was estimated at about 1000 mud loads for the *cliff swallow (*Hirundo (Petrochilidon) pyrrhonota*) from an examination of the nest external surface (Emlen 1954), but actually counting the number of collection trips gives a figure of 2500 for the completion of the less complex nest of the house martin (*Delichon urbica*) (McNeil & Clark 1977).

It could be argued that this method of building is not strictly moulding because the shape of the nest results from the appropriate placing of each succeeding mud pellet rather than from the

subsequent shaping. Nevertheless, moulding behaviour is then
needed to fuse the loads together to create a coherent wall. This
poses the birds the same two problems that face a potter building a
pellet or coil pot. The first of these is the trapping of air between
successive loads; the second is the difference in moisture content
between the pellet being applied and those already incorporated.
Both lead to a weakened structure: air spaces are, in effect, cracks,
and mud contracts on drying, causing cracks to develop where
patches of nest wall of different water content lie side by side.

Combating these problems explains the 'vibrating' movement of
the head by *cliff swallows when applying a new mud load to the
nest (Emlen 1954), and also of loading mud first in the beak and
then on top of it by the same species (Samuel 1971). The purpose of
the latter behaviour is not entirely obvious, but a similar behaviour
is shown by the barn swallow (*Hirundo rustica*), which appears to
place more liquid mud on top of the beak. Consequently, when the
new pellet is pressed against the existing nest, the moister mud on
top of the beak forms a bridge between the two, avoiding cracks
developing when the nest dries. The vibration of each new load
probably causes a partial liquefaction of the mud through the

phenomenon of thixotropy. This disperses the moisture more evenly and allows the fresh mud to overrun small air spaces. This
technique is also known for mud dauber wasps (*Sceliphron*,
Sphecidae), which make a buzzing noise when applying fresh mud
loads during brood cell construction (Hansell 1984).

It is remarkable that the mud nests of house martins and *cliff
swallows are built on sites where none of their weight is supported
from below, but the inclusion of some grasses or horse hair into the
mud may give additional tensile strength. These materials are
brought separately and placed on the mud, becoming embedded in
it as the nest wall grows (Emlen 1954, Samuel 1971). A second
danger to a nest attached in this way is that it becomes distorted
under its own weight before the mud dries. This is avoided through
bouts of building interrupted by pauses for the new portion of wall
to dry out (Emlen 1954). This pattern of construction is sometimes
revealed in the banded appearance of the nest, showing that mud
was gathered from slightly different locations in different collecting
bouts (Emlen 1954).

There are bird species that build cup nests supported from below
and incorporating a greater proportion of vegetation in the mud.
However, these nests show that similar construction rules apply as
to the wall-attached nests of house martins. The *white-winged
chough (*Corcorax melanorhamphos*), apostlebird (*Struthidea
cinerea*) and *magpie-lark (*Grallina cyanoleuca*) vibrate loads of
mud-impregnated vegetation as they apply them to the nest rim
(Rowley 1970), and may pause between building bouts (Rowley
1978).

Several other species from various families build vegetation nests
to which mud is added to line the cup. Rowley (1970) describes this
as *plastering*. The smooth finish of this workmanship in species
such as the black-billed magpie (*Pica pica*) or song thrush (*Turdus
philomelos*) is probably the result of *shaping* movements of the
breast and feet like those seen in species using other construction
methods.

4.5 Piling up

The simple *piling up* technique of the magpie goose (*Anseranas
semipalmata*) is the bending down of rooted rush stems towards the
middle of the nest site to form a radial array (Davies 1962, Fig. 4.4).
Typically, however, the technique involves the construction of a
twig platform in a tree, with a more or less distinct cup. It is seen in
several families, e.g. storks (Cicioniidae), pelicans (Pelicanidae),

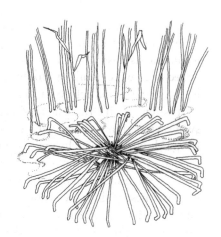

Figure 4.4
The magpie goose
(*Anseranas semipalmata*)
constructs its nest either
by bending down rush
stems one at a time held
in the beak or by pushing
down several stems held
in the crook of the neck.
In this way it builds up a
nest platform above the
water. (Adapted from
Davies 1962.)

herons (Ardeidae), eagles (Accipitridae), crows (Corvidae) and pigeons (Columbidae) (Fig. 4.5). Strictly this is a fetch and drop technique, as described for the little sparrowhawk (*Accipiter minullus*) (Liversidge 1962) and no attempt at interlocking one twig with another is made; however, some degree of interlocking of twigs may be present in species belonging to all these families. Of the 300 or so species of pigeon and dove, more than half build platform nests in trees (Goodwin 1983). Some of these nests are composed of a few twigs only, that of the brown-backed emerald dove (*Chalcophaps stephani*), for example, while others, such as that of the Marquesan imperial pigeon (*Ducula galeata*), are much more substantial. All, however, suggest construction by piling up. Careful examination of the moderately sturdy nest of the *woodpigeon (*Columba palumbus*), on the other hand, shows that the projecting ends of twigs are, in fact, bent and twisted into the body of the platform to give it additional rigidity (personal observation). In this species also, as in some other European *Columba* species, the droppings of the nestlings impregnate and dry within the platform, possibly reinforcing it as the weight of the chicks increases. By contrast, in some tropical New World genera (*Zenaida*, *Leptotila* and *Geotrygon*), chick faeces are deliberately removed by the parents (Goodwin 1983).

Construction of the nest of the black-billed magpie *Pica pica* certainly involves more than simple, sequential placement of twigs to create the roofed cup nest. An average period for the manipulation of a twig at the nest is about 25 seconds (Kerr & Hansell, unpublished). Manipulation includes moving the twig from position to position and *shuddering*, in which the body is shaken rapidly from side to side or vertically (Goodwin 1976), apparently in an effort to entangle the twig more securely among the projections of

Figure 4.5
The nest of the *spotted
dove (*Streptopelia
chinensis*, Columbidae) is
a platform of twigs laid
across one another.
(Photograph by M.
Hansell.)

the others. Similar manipulation of twigs is shown by *rooks
(*Corvus frugilegus*) (Rutnagur 1990). It may therefore be that construction of a stick platform, with or without a deep cup, should
not strictly be placed under the heading of piling up, because an
element of intertwining is involved as well as behaviour required in
shaping the cup. Nest construction of the white-tipped brown jay
(*Psiloirhinus mexicanus*) consists of building up a platform of
twigs on fine branches at the end of a long horizontal limb, but a
depression is shaped in the middle of the platform using the breast
and scrabbling with the feet (Skutch 1960).

For nests built on the ground, there is less need for rigidity of
structure, so the reed platform of the magpie goose is adequate. The
nests of waterfowl generally, including some other ground-nesting
Anatidae like the *black scoter (*Melanitta nigra*), also appear to be
little more than a haphazard pile of grass stems and down feathers
which the duck places round her to build up a cup rim. The cup is
shaped with movements of the feet and breast and lined with
feathers plucked from the bird's own breast and placed in the cup
under her (Kear 1970). For Rallidae, however, which nest on the
ground or water using long stems or leaves of aquatic vegetation,

the strips of material are bent around to form a cup and projecting pieces are bent inwards or across one another to create an inter-twined texture, e.g. the *Virginia rail (*Rallus limicola*), or the *white-throated crake (*Laterullus albigularis*).

4.6 Sticking together

The mud users described under *moulding* grade into species which can be said to use the mud to stick together the vegetation compo-nents. The white-winged chough and apostlebird have already been mentioned, and Rowley (1970), in his compilation of mud-using birds, adds species like the black-legged kittiwake (*Rissa tridactyla*) and the *eastern phoebe (*Sayornis phoebe*, Tyrannidae), whose mud, moss and rootlet nest is sited like a rather bulky barn swallow, nest on a ledge or wall (Weeks 1977).

Profiled museum specimens include some New World Fringil-lidae like the *common grackle (*Quiscalus quiscula*), which re-inforces its grassy cup with mud, and the *spotted morning-thrush (*Cichladusa guttata*, Muscicapidae), with an even more muddy texture and a neat cow dung lip to the cup. More daring designs are found among the versatile Furnariidae, for example the rootlet, grass, moss and mud nest of the *sharp-tailed streamcreeper (*Loch-mias nematura*) suspended from vine stems over water (110g dry weight for a nest build by a 25 g bird), or the nest of the *wren-like rushbird (*Phleocryptes melanops*), a domed, fibrous nest sited on vertical reed stems and impregnated with mud to give it the solidity of a coconut.

Some swifts (Apodidae) also reinforce vegetation nests with mud. The *chestnut-collared swift (*Cypseloides rutilus*) builds a bulky moss and mud nest in soaking wet sites near to running water. The nest is, in fact, a living structure, the moss continuing to grow, rooting the nest to the substrate (*Museum nest collection notes*). The swifts also provide the examples of vegetation stuck together with mucus like the shallow, rock crevice nest of the *white-throated swift (*Aeronautes saxatalis*) of plant down and feathers cemented together with mucus. Convergently, the cliff flycatcher (*Hirundinea ferruginea*, Tyrannidae) also builds a nest of roots and straw held together with a salivary mucous adhesive (Sick 1993). The most architecturally satisfying example of the sticking together technique is probably that of the *chimney swift (*Chaetura pelagica*, Apodidae), a wall-attached bracket of straight twigs held together with salivary mucus (Pearson & Burroughs 1936).

4.7 Interlocking

It is fundamental to this class of construction that the materials are all dry and therefore hold together without the aid of any adhesive. Three rather different construction methods are recognised within this: *entangle, stitches* or *pop-rivets* and *velcro*. The second of these is exhibited by a select minority of species, but the other two are important for being much more pervasive. Interlocking is possibly the most important category of nest construction methods and, fortunately, we have enough behavioural information to understand the basics of how birds accomplish it. This group of builders, in fact, illustrates most effectively the principle that nest materials stay together both because of the spatial relationship established between them through the behaviour of the builder, and because the materials have properties which hold them in that relationship. When these dual principles are seen time and again it must be concluded that materials are chosen with care to create a strong nest and frequently to make construction itself more easy.

a) Entangle

It has already emerged when considering twig nests built by *piling up* that some care may be taken in manipulating a new piece so that it fits effectively into place. The slaty spinetail (castlebuilder) (*Synallaxis brachyura*), which builds a domed twig nest complete with entrance tube, is recorded by Skutch (1969) as pulling and pushing the twigs until they 'fitted closely together into a coherent whole'.

In nests composed of grasses or other similarly flexible vegetation strands, some element of intertwining can usually be recognised. The screaming piha (*Lipaugus vociferans*) makes a light, open framework of loosely interlaced twigs (Sick 1993), and the nest of the *striped tit-babbler (*Macronous gularis*, Sylviidae) is composed of broad strips of monocot leaf simply but neatly tucked into one another (Fig. 4.6). Aichorn (1989) emphasises the simplicity of the building of the grass and rootlet cup nest of the white-winged snowfinch (*Montifringilla nivalis*), which inserts materials into the nest with vibrations of the beak and shapes the cup with movements of beak and feet. To start its suspended, retort-shaped nest of fibrous rootlets and thread-like fungal hyphae, the female snowfinch entangles the thin fibres round a twig. Additional strands are then worked into the suspended mass until it grows into a loosely tangled ball. The fabric is thickened to from a nest cavity as she *tucks in* loose material with rapid pushes of the beak. Skutch

Figure 4.6
The nest of the *striped tit-babbler (*Macronus gularis*, Sylviidae) is made of broad strips of monocot leaf entangled in one another by a series of neat tucks. (Photograph by M. Hansell.)

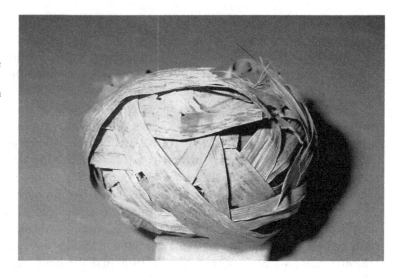

(1960) calls this tucking in behaviour 'matting' or 'felting' and describes the behaviour of a highly effective exponent of it, the sulphur-rumped flycatcher (*Myiobius barbatus*). The completed nest of this species is a domed basket with an entrance porch, hanging from a slender vine. The female begins it by binding strands of vegetation to the suspension with quick revolutions of her whole body. On creating a loose mass of hanging material, she works from within it to create a cavity into which she carries most of the subsequent loads of fibrous plant materials. These are then worked into the existing fabric as the bird continues to stretch the wall from within, finally producing an ample cavity with a densely matted wall. Binding material to the nest site may be the characteristic method of starting the nest when the entangling technique is employed. The white-winged becard (*Pachyramphus polychopterus*) commences its nest by securing strips of bark with complete turns of the material round the supporting branches (Skutch 1969).

In the above examples it is not obvious that the properties of the materials themselves combine with the probing and binding behaviour, entangling the materials to give the nest its integrity. For construction with vine tendrils this is quite evident. Skutch (1969) describes the behaviour of the *rufous piha (*Lipaugus unirufus*) (Fig. 4.7) early in the construction of its cat's cradle of a vine tendril nest:

> The Piha placed her latent contribution on her incipient nest, then remained sitting on it, pulling up the projecting ends or loops of tendrils and tucking them into the tangle beneath her. Stiff and wiry

they tended to spring outward again, but being birds of great persist-
ence, she forced some to remain as she desired. While engaged in this
task, she turned sideways from time to time, slowly revolving on her
nest and performing the same operation on all sides.

Nests made with vine tendrils are not numerous. The nests of six
species, each from a different family, were recorded in the nest
survey, demonstrating the versatility of the material and effectiveness of the technique. They were: the minimal open-work platform
of the *rufous piha (Tyrannidae) (Fig. 4.7), the *wompoo fruit-
dove (*Ptilinopus magnificus*, Columbidae), the *rusty pitohui
(*Pitohui ferrugineus*, Corvidae), the *stripe-headed tanager (*Spin-
dalis zena*, Fringillidae), the *hwamei (*Garrulax canorus*, Sylviidae)
and the massive, hanging roofed basket of the *metallic starling
(*Aplonis metallica*) (dry weight of 450 g).

Construction with plant materials, whether grass, bark or vine
tendrils, requires not only interlocking with the beak but, as the
materials accumulate, the shaping of a nest cavity to hold the eggs.
This, again, is achieved mainly with the breast and legs. The shaping behaviour cannot be observed directly in domed nests, although
Skutch (1960) describes the enlargement of the hanging nest chamber of the *sulphury flat-bill (*Tolmomyias sulphurescens*) when she
takes material into the chamber: 'the thin fabric that surrounds her
bulges and trembles as, with vigorous movements of feet and wings,
she shapes the chamber and expands it to the proper size'. Shaping
with feet and wings is also recorded for open cup builders, for
example the blue-black grosbeak (*Cyanocompsa cyanoides*)
(Skutch 1954).

b) Stitches and pop-rivets

The *common tailorbird (*Orthotomus sutorius*, Sylviidae) forms its
purse nest by linking together the margins of green leaves attached
to a shrub or bush by means of fibrous stitches. I have identified in
the coarse yarn that forms the stitches, lepidopteran silk, spider silk
and plant down. The stitches are made by driving the thread
through the leaf, grasping it on the other side and driving it through
again. The coarseness of the thread and the elasticity of the green
leaf springing back to grip the thread passing through the hole
prevent the stitches from unravelling. An almost identical fastening
with a similarly mixed yarn is used by the *grey-crowned prinia
(*Prinia cinereocapilla*) and *golden-headed cisticola (*Cisticola
exilis*) which, although more traditionally placed in the family
Sylviidae, are given a separate family (Cisticolidae) by Sibley and

Figure 4.7
The nest of the *rufous
piha (*Lipaugus unirufus*,
Tyrannidae) is an
open-work platform of
stiff, interlocking vine
tendrils. (Photograph by
M. Hansell.)

Monroe (1990). In the same family, the *grey-backed camaroptera
(*Camaroptera brevicaudata*) uses a combination of stitches and
scores of pop-rivets to hold together its leaf purse nest. The pop-
rivets are tufts of the plant down lining of the nest, driven through
the leaf surface (Fig. 4.8).

Indisputably in another family, the Nectariniidae, are the *little
spiderhunter (*Arachnothera longirostra*) and long-billed spiderhunter (*Arachnothera robusta*), which suspend a nest chamber complete with entrance tunnel below a large leaf by means of about 150
fine threads (Madge 1970), each secured by a pop-rivet driven from
below through the leaf membrane. The threads are silk, thickened
at the end into a distinct knob which, on being driven through the
leaf membrane, is trapped by the recoil of the leaf tissue (Fig. 4.9).

c) **Velcro**

The most effective employment of the interlocking technique is
probably the one which is least appreciated. It is the entanglement
of vegetation in threads of silk. This technique I feel justified in
calling *velcro* fastening after the garment fastening Velcro invented

Figure 4.8
The *grey-backed
cameroptera
(*Cameroptera
brevicaudata*) attaches the
inner nest lining to the
outer envelope of growing
leaves by driving through
the leaf membrane dozens
of plant-down pop-rivets.
(Photograph by M.
Hansell.)

by George de Mestral, a Swiss engineer, who in 1940 was inspired by the sight of plant burrs becoming entangled in the fur of his dog (Jacobs 1996).

My own nest examinations show that some species from 25 of the 45 (56%) passerine families (Sibley & Monroe 1990) use silk as a structural material and, in the non-passerine family the humming-birds (with about 300 species), silk use may be almost universal (Table 4.1). The silk samples in a survey of the nests of 110 genera from six families show that very few of these use silk as an adhesive. The only adhesive silk available to birds is the web silk of ecribellate spiders, which use sticky droplets to capture flying insects by adhesion. Many of these species produces orb webs, which we admire for the small amount of silk that can be used to produce such a large capture surface. For birds, such webs are not very valuable; instead, they collect thick sheets or blobs of silk from sheet webs, tubular retreats and spider egg cocoons, or from lepidopteran pupal cocoons. All these have dry threads and so can only be used in nest building as the looped material of a *velcro* fabric.

Not all plant materials will provide the numerous projections

Figure 4.9
The head of a silken
pop-rivet, one of many
driven through the leaf
from below to secure the
hanging nest of the *little
spiderhunter
(*Arachnothera
longirostra*). (Scale:
separation of two white
bars is 100 μm.)
(Photograph by Margaret
Mullin.)

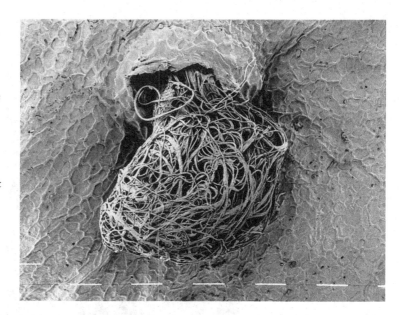

that can engage the silken loops, so it can be predicted that silk users
will select the vegetation component of their nests to form an
effective partnership. For members of the Aegithalidae, this seems
to be the case. The nest of the *long-tailed tit (*Aegithalos caudatus*)
contains the silk of more than 600 spider egg cocoons of a fluffy
texture that provides lots of loops. In Britain these are then generally entangled with sprigs of the small-leafed moss *Eurynchium
praelongum* or another species with similar branching morphology
(Fig. 4.10, Hansell, in preparation). This suggests that *A. caudatus*
may be rather specialised in its selection of moss, although in silk
builders as a whole, various plant materials are found in combination with spider silk (see Section 5.4c).

The virtue of *velcro* as a fastening is twofold. Firstly, the
materials can be locked together with only simple movements and,
secondly, they can be repeatedly separated and refastened. This
should allow the construction behaviour of *velcro* builders to be
simple. In the case of the building of the elegant, complex nest of the
*long-tailed tit this proves to be the case. Tinbergen (1953) de-
scribes a *long-tailed tit pair collecting moss, which is pressed into
the fork of a branch with quivering beak movements. Spider cocoons are then applied to the nest platform where they stick 'at
once'. The bird can then draw a thin strand from a cocoon, which is
pressed down at another spot. So, silk is fastened to branch and
moss to create a nest foundation. The gathering of more silk and
moss, with the employment of *right–left* or *vertical* movements of

Table 4.1. **A compilation of the 26 families of birds for which there is evidence of some species using arthropod silk in nest construction**

Non-passerine families	
Trochilidae	

Passerine families	
Thamnophilidae	Paridae
Tyrannidae	Aegithalidae
Maluridae	Regulidae
Meliphagidae	Pycnonotidae
Pardalotidae	Cisticolidae
Eopsaltriidae	Zosteropidae
Irenidae	Sylviidae
Laniidae	Nectariniidae
Vireonidae	Melanocharatidae
Corvidae	Paramythiidae
Bombycillidae	Passeridae
Muscicapidae	Fringillidae
Certhiidae	

In addition to the nest survey, evidence of silk use was obtained from the following: Chapin (1953), 1954), Coates (1990), Falla *et al.* (1989), Harrison (1975, 1978), Madge (1970), North (1901–4), Oliver (1955), Sick (1993), Skutch (1976), Smythies (1986), Snow (1976), Van Someren (1956).

the beak to draw out and fasten the silk, builds up the nest wall and finally the domed roof. The completion of the nest requires other phases: the shaping of the cavity, application of an exterior lichen layer and insertion of a feather lining, but Tinbergen's conclusion is perceptive: 'the most amazing thing about it (*the construction behaviour*) is, in my opinion, the fact that so few, so simple and so rigid movements together lead to the construction of so superb a result'.

Specific choice of materials and the refastening properties of *velcro* are illustrated in a fellow aegithalid, the *bushtit (Psaltriparus minimus*). The nest starts as a shallow pouch fastened to a forked twig with spider silk. Adding material to the inside, the birds stretch the fabric with vigorous movements, eventually extending it to form a deep bag. This huge expansion is dependent on the ability of the *velcro* fabric to release and reform attachments.

Figure 4.10
Sample of the nest wall of
the *long-tailed tit
(*Aegithalos caudatus*)
showing loops of spider
cocoon silk around the
small leaflets of moss.
This combination builds
up a wall held together by
the *velcro* principle.
(Scale: separation of two
white bars is 100 μm.)
(Photograph by Margaret
Mullin.)

Addicott (1938) describes the material of the *bushtit nest as
spider web, lichens, mosses, grasses, and flowers of the oak *Quer-
cus agrifolia*, which are all 'woven together'. My examination
shows that this composite is a *velcro* textile and that, significantly,
the lichen is of a particular stellate morphology with long, stiff
projecting spines (Fig. 4.11). I also sampled lichen of this morphology from the nest of the *bran-coloured flycatcher
(*Myiophobus fasciatus*, Tyrannidae). The specificity of the choice
of this lichen is confirmed by Skutch (1960), who describes the
nest of the black-eared bushtit (*P. melanotis*, Aegithalidae) (syn-
onymous with *P. minimus* in Sibley & Munroe 1990) as a very
open, extremely frail fabric composed of 'finely branched, gray
foliaceous lichens and held together by cobweb, aided no doubt by
the fine, ciliate projections from the lichen itself; only one kind of
lichen, a species abundant on the bark of trees in Guatemala, had
been employed'.

Using silk as the looped material to ensnare vegetation over-
comes the difficulty posed by other building techniques of getting
the nest started. Skutch (1954) describes the female *white-
collared seedeater (*Sporophila torqeola*, Fringillidae) as starting
nest building by wrapping strands of cobweb around the bran-
chlets. Standing in the centre of the nest site, she '...soon has the
entire nest outlined in cobweb'. This, Skutch comments, with a bill
'little suited to manipulating such delicate material'. A seemingly

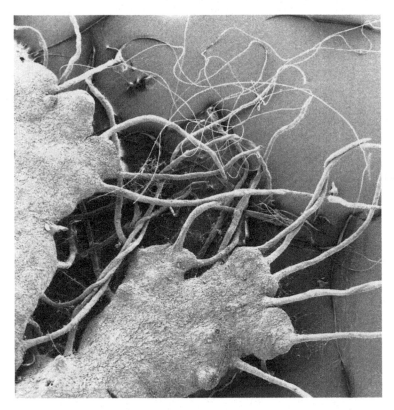

Figure 4.11
The nest of the *bushtit
(*Psaltriparus minimus*) is
a *velcro* fabric featuring a
characteristic lichen,
which bears stiff
projections, shown
entangled with threads of
spider cocoon silk.
(Photograph by Margaret
Mullin.)

identical method of starting the nest is employed by the *common
goldcrest (*Regulus regulus*). Silk is wrapped round twigs at the nest
site, pulled out to a thread and wrapped at a second point, building
up a scaffolding to provide for the addition of a hanging mossy cup
(Thaler 1976). The construction of the nest cup is accompanied by
further stretching and wrapping of silk before the shaping of the
next cup with energetic movements of the breast, wings and feet
(Fig. 4.12).

The cup-shaping movements of *velcro* textile nests seem in es-
sence to be the same as those already reported for builders using
the interlocking technique. The bellicose elaenia (*Elaenia
chiriquensis*, Tyrannidae), which starts its nests by attaching twigs
to the supporting branches, completes the nest by pressing her
breast down into the nest cup and then outward with rapid backward pushes of the feet. Very similar shaping movements are
shown by another elaenia species, *E. flavogaster*, by the yellow-
green vireo (*Vireo flavoviridis*, Vireonidae) and by the tropical
gnatcatcher (*Polioptila plumbea*, Sylviidae) (Skutch 1960).

Figure 4.12
Nest construction by the
*goldcrest (*Regulus
regulus*) begins with the
creation of a framework
of spider silk to contain
and hold the mossy
vegetation (a to d). The
cup is finally shaped by
vigorous breast
movements and scraping
with the feet (e). (Adapted
from Thaler 1976.)

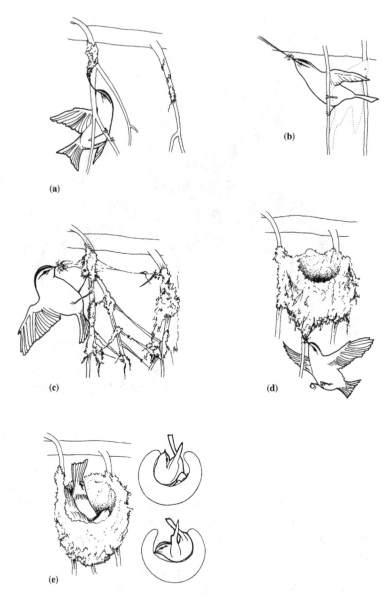

(a)

(b)

(c)

(d)

(e)

4.8 Weaving

Getting a nest started is a particular problem for birds because the
material has to be fitted to an unpredictable landscape rather than a
regular structure created by the builder (see Chapter 5). For
weavers the task is especially difficult; firstly, woven structures bear
loads in tension and so the first strips must at least bear their own
weight and, secondly, the strips have no inherent properties to

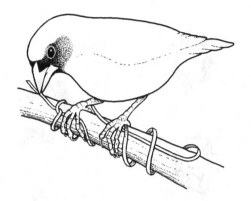

Figure 4.13
The assistance of both feet while knotting a grass strand helps a male red-billed weaver (*Quelea quelea*) to begin weaving a nest. (Adapted from Chapin 1954.)

secure them until tied to branches from which the nest will be suspended. Spiral wrapping round an attachment twig may give a strand temporary stability, but it must be secured with a hitch or knot; this may require integration of movements of beak and feet as shown by the male red-billed weaver (*Quelea quelea*) (Fig. 4.13, Friedmann 1922, Chapin 1954). The same behaviour of spiral binding and hitch fastening is also characteristic of the male *masked weaver (*Ploceus velatus*) (Howman & Begg 1983, 1995).

The key elements of building up the nest wall are passing the vegetation strand under itself and pulling through, and of progressing the same strand through the nest by repeatedly passing it in, grasping the end and pulling it through. This is seen in the sequence in which the *village weaver (*Ploceus cucullatus*) incorporates a leaf strip into the ring of material it builds up below the nest suspension (Fig. 4.14, Collias & Collias 1962). Similar patterns can be recognised from examination of completed nests of other Old World weaver species (Ploceinae) (Collias & Collias 1964a).

Another group of birds generally credited as weavers are the New World oropendolas, caciques and orioles (Icterini, Emberizinae, family Fringillidae). The birds themselves are, however, strikingly different from the ploceine weavers. Their beaks are different in shape, it is the female who builds and, in particular, they are normally much larger birds, as much as ten times the weight of an average ploceine (Fig. 4.15). However, in building technique these Old and New World weavers are markedly convergent.

The female *Montezuma oropendola (*Gymnostinops montezuma*) begins by tearing strips from a monocot leaf such as a banana by making a nick in the leaf and then flying off, tearing a fine strip of fresh green material as it departs (Skutch 1954). For the male village weaver the same bite-and-pull technique is performed on the leaves of elephant grass (Collias & Collias 1964a). The

Figure 4.14
To weave a single strip of
elephant grass into the
nest requires the male
*village weaver (*Ploceus
cucullatus*) to repeatedly
press the same strand
through the nest, grasp it
again and pull it through.
(Adapted from Collias &
Collias 1962.)

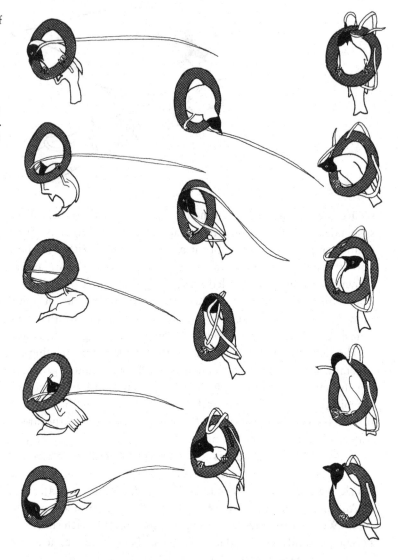

oropendola secures the long, fine strips of green leaf round a twig
'pushing the fibres under the twig and pulling them over, intertwining and knotting them carefully' (Skutch 1954). In this way a ring of
materials is formed, reminiscent of that of the village weaver. She
now leans down inside the ring, extending it downwards to form a
tube into which she then disappears to continue her work on the
inside, eventually closing it at the bottom to create the nest chamber. The sequence of construction of the yellow-rumped cacique
(*Cacicus cela*) is apparently very much the same (Skutch 1954).

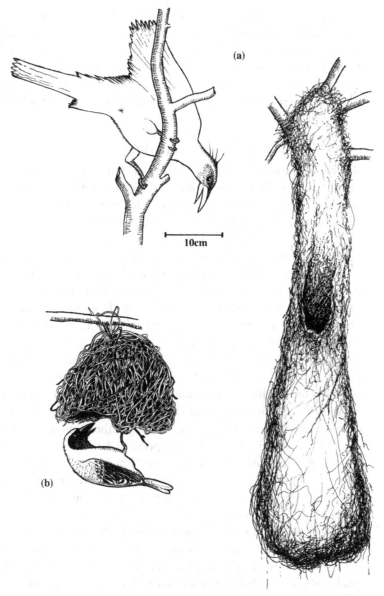

Figure 4.15
The weaving technique of nest building has become highly developed in two different groups of birds. (a) In the New World oropendolas and caciques, nests are constructed generally by females weighing between 50 and over 200 g, and the nests are in the form of a deep purse with an entrance near the top. (b) The largely African weavers, are much smaller birds, of 20–40 g; nests have a side or downward directed entrance, and it is normally the male that does the majority of building.

Examination of the fabric of completed nests of the *yellow-rumped cacique and the *crested oropendola (*Psarocolius de-cumanus*) confirms the presence of hitches and spiral binding. It also reveals that individual strips are embedded in the nest fabric either by being repeatedly inserted and pulled through at one end or, more usually in the cacique, by being held in the mid portion, inserted as a *loop tuck* which is then drawn into the nest wall.

In both the crested oropendola and the yellow oriole (*Icterus nigrogularis*) single strands are woven into the nest wall following a complex path. Loops, hitches, loop tucks and spiral coils are found in both, although complex workmanship and fine spiral coils at the nest rim are particularly characteristic of the oropendola (Heath & Hansell 2000). So, most of the techniques shown by ploceine weavers (Fig. 4.16) are also carried out by oropendolas and caciques (Icterini) to a high standard, in spite of the size of the birds.

4.9 How difficult is nest building?

Reviewing the nest building of birds, some important trends can be identified. The first is that birds do not have anatomy that is specialised for the building technique each shows. There is a lack of evidence here but, whereas from a bird's beak you might reasonably deduce its diet, you would have difficulty in predicting the nature of its nest. Building techniques must therefore be strongly dependent upon behaviour yet, in spite of this, there is little evidence of behavioural complexity. Instead, some rather similar movements are employed by species using different construction principles; for example, the wrapping or binding movements to create nest attachment points, the pushing of new material into the existing fabric with the beak, or the shaping of the cup with movements of the breast and feet. These suggest that nests can be constructed by whatever principle using a fairly limited repertoire of stereotyped movements, exactly as Tinbergen (1953) concludes for the construction of the architecturally rather complex nest of the *long-tailed tit. We know of specialised brain structures in birds concerned with song production (Nottebohm 1980) and for remembering sites of cached food (Krebs 1990); the day that specialised brain areas for nest building are identified, I will need to reassess this conclusion.

Stereotyped, repeated movements and simple building rules can produce an elegantly simple or sophisticated structure when carried out on standardised building materials. This is the principle of the brick wall. So, much depends on the careful choice or manufacture of building units and materials. Evidence for this is seen in the specific choice of mud by cliff and barn swallows (Kilgore & Knudsen 1977) and of green leaf strips by the village weaver (Collias & Collias 1964a). It is also evident in species using the interlocking principle, in particular those that select *velcro* materials. Velcro is now the preferred fastening for children's clothing, because the attachment principle is in the material not in the behaviour. By contrast, tying shoelaces is a landmark in a child's development. This is hardly a scientific case, yet it is illustrative of the view that

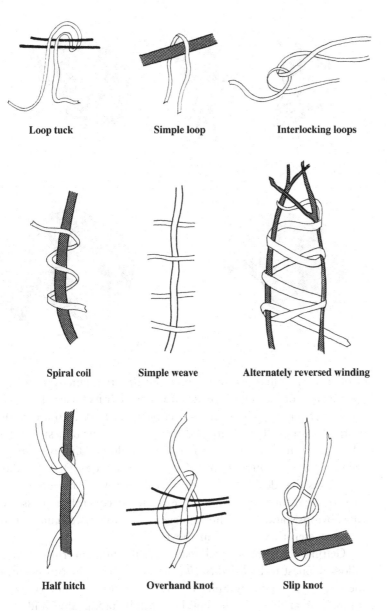

Loop tuck   **Simple loop**   **Interlocking loops**

Spiral coil   **Simple weave**   **Alternately reversed winding**

Half hitch   **Overhand knot**   **Slip knot**

Figure 4.16
Various common stitches and fastenings used by weavers (Ploceinae). (Adapted from Collias & Collias 1964b.)

weaving is the most difficult nest building technique for birds (Howman & Begg 1995). Consequently, if any birds show development of building skills through experience, then they are most likely to be weaverbirds.

A nesting bird must build a wall round itself (*fit the structure to itself*), in the case of a cup nest, first below the builder and then on all sides of it. A standard reaching-out movement to fit each piece of material defines both the size of the nest and its perfect circularity.

Figure 4.17
The construction of the
nest chamber and
entrance porch by the
*village weaver (*Ploceus
cucullatus*) illustrates how
the dimensions of the nest
can be defined simply in
terms of the birds reach
when standing on the nest
ring. (Adapted from
Collias & Collias 1962.)

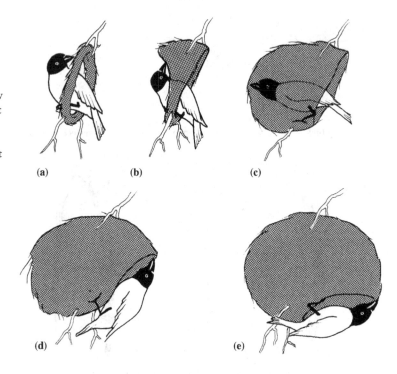

(a) (b) (c)

(d) (e)

Collias and Collias (1962) pointed out that the dimensions of the
domed nest of the village weaver are defined by its forward reach to
construct the nest chamber and its backward reach to create the
entrance porch while positioned in the original ring of nest material
(Fig. 4.17). It is time to see if such a simple building principle is
general to the construction of all or most nests. The building
process is that described earlier (Section 4.7a) for the very simple
nest of the rufous piha (Skutch 1969) and is frequently expressed in
the obvious spiral or swirling appearance of completed cup nests of
many passerines viewed from above.

Getting the nest started may require different rules, which
allow the nest to be *fitted to the landscape*. The obvious problem
here is that the topography of the nest site or geometry of its
branch arrangement will not be entirely predictable and may
therefore require greater flexibility in the behaviour than continuing construction after a nest has been established. Nickell (1958)
pointed out that the hanging nests of, for example, *Baltimore
(northern) oriole (*Icterus galbula*) and *red-winged blackbird
(*Agelaius phoeniceus*) show greatest variability in their design in
the attachment (Fig. 4.18), confirming the more varied behaviour
needed at this stage. The added problem of getting started is that
of not having other nest materials to which to attach the current

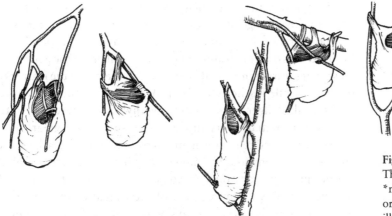

Figure 4.18
The suspended nest of the
*northern (Baltimore)
oriole (*Icterus galbula*)
illustrates how variability
in nest design may result
largely from differences in
the configuration of the
chosen nest site. (Adapted
from Nickell 1958.)

beak load. This can be seen in the number of twigs dropped, for example, by rooks and magpies trying to create the initial nest platform. This contrasts with the extremely rugged structures they eventually create. By comparison, the silk users have an especially easy task, because this material attaches as readily to supporting twigs as to other nest materials, whereas weavers have a particularly difficult task because the first attachments require the most secure knots and hitches. In circumstances like these, skill learning might be beneficial.

The instinctive nature of bird nest building has long been a point of controversy. Herrick (1911) is refreshingly sceptical about the evidence for learning, stressing the need for experimental evidence, and critical of Alfred Russell Wallace (1867) for comparing birds with humans in their capacity to build. Surprisingly, a deprivation experiment may have been conducted as early as the seventeenth century by the remarkable scholar and naturalist John Ray, who writes of the species-typical nest architecture of birds and concludes 'and this they do though they never saw nor could see, any nest made, that is though taken out of the nest and brought up by hand' (Raven 1950). In spite of this, evidence on the genetic basis of any building behaviour is rare. The offspring of both deermice (*Peromyscus maniculatus*) and Oldfield mice (*P. polionotus*), reared in plastic laboratory cages for 20 generations, produced species-typical burrows. Hybrids between them dug burrows characteristic in all details of *P. polionotus*. These F1 hybrids backcrossed to the recessive *P. maniculatus* showed varying mixtures of burrow characters of the two species (Dawson, Lake & Schumpert 1988). Burrow architecture is, it seems, determined by a number of loci at which *P. polionotus* alleles are dominant.

The hybridisation approach has been applied to bird nest building in the lovebird genus *Agapornis*. These are all cavity nesters, but collect strips of vegetation to line the cavity. The peach-faced lovebird (*A. roseicollis*) and some other members of the genus carry these strips, rather improbably, by tucking them into the feathers on the back, while Fischer's lovebird (*A. fischeri*) carries single pieces in the beak. Dilger (1962) found the hybrids between these two species were initially very disorganised, spending a great deal of time inserting and removing material from their back feathers, their efforts almost invariably ending in failure. The sterility of the hybrids prevented any further genetic analysis but, interestingly, by the end of the two months, 41% of carrying attempts were in the beak, although these were always preceded by some tucking movements. Further improvements were evident at the end of two years. So, although the building behaviour was simple, there was evidence of improvement with experience. On the other hand, Sargent (1965), using zebra finches (*Taeniopygia guttata*) raised in experimentally altered nests to test Wallace's (1867) suggestion of learning nest architecture as a chick, found only slight effects.

Collias and Collias (1984) comment on the lack of experimental studies on the ontogeny of bird nest building. Their own work on the village weaver remains today, a notable exception, although we do have anecdotal evidence of an extended period of learning construction techniques in male bowerbirds (Ptilonorhynchidae) (see Chapter 8). Young, male village weavers, aviary-reared with or without experience of green building materials, showed a preference for green when offered a range of colours of materials for their first nest building attempt, but the degree of preference of the inexperienced birds became more marked over the first three days (Collias & Collias 1964a). A preference for flexible over rigid building materials was also common to experienced and deprived birds, but naive yearlings, trying to tear strips of reed grass, were initially inept, whereas experienced birds of the same age were efficient. In the development of weaving skills, birds obtaining more practice showed greater proficiency. Naive yearling males were unable to weave a single strip in the first week that reed grass was made available, whereas the experienced controls were quite proficient. Even after three weeks of practice, birds of the deprived group were only capable of weaving 26% of strips compared to 62% by the controls (Collias & Collias 1973).

So, although some elements of village weaver nest building were unaffected by these deprivations, others were clearly impaired.

Improvement in performance seems to continue beyond the first year; in the wild, first-year males build looser, more untidy nests than older males (Collias & Collias 1964a). What is needed now is a detailed video analysis of the ontogeny of building behaviour and an experimental investigation of how the various weaving actions are affected by trial and error or even observational learning. The most exciting models for this would be the New World Icterini rather than the Old World Ploceinae because their size should be pushing the weaving technology closer to its limits.

4.10 Tool use and tool making

It would be wrong to omit from this book any consideration of the making and use of tools by birds. Not that tool use is claimed to be involved in nest building, but, more extraordinarily, because it is treated as if it is a quite unrelated subject. Both tool making and building behaviour involve manipulation of the environment, yet the former is often regarded as an indication of advanced mental processes, the latter interesting but largely under genetic control. It is important to understand the reasons for these marked differences in attitude. In this respect the description of two foraging tools manufactured by the New Caledonian crow (*Corvus moneduloides*) (Hunt 1996) provides a very useful case study because it clearly lays out the argument supporting the special status of tool making. The two tools in this case are a twig cut at the base to leave a recurved spur, and a tapered strip of *Pandanus* leaf, the surface of which is covered with fine teeth (Fig. 4.19). These tools are claimed to be important because they exhibit: (1) a high degree of standardisation; (2) distinctly discrete tool types shaped to a particular role; and (3) the use of hooks (to extract insect prey from cavities in the wood). Such features, it is claimed, 'only first appeared in the stone and bone tool-using cultures of early humans after the Lower Paleolithic, which indicates that crows have achieved a considerable technical capability in their tool manufacture and use.'

I have emphasised in this chapter that the great variety of sophistication of bird nest construction is apparently achieved with a small repertoire of rather stereotyped behaviours. Learning is involved for the handling of some materials and may be fairly general; however, the learning ability of birds in other contexts is not in doubt. So, how is it that so much significance can be attached to the shaping of a twig or piece of leaf?

Figure 4.19
Two types of foraging
tool used by the New
Caledonian crow (*Corvus
moneduloides*): (a) a
narrow stick, and (b) a
tapered palm leaf with
serrated edge. (Adapted
from Hunt 1996.)

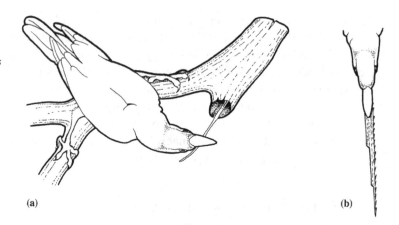

(a)   (b)

Take first the claim of a *high degree of standardisation*. This is, I
have argued, just what you would expect to make any construction
behaviour easy and the end result simple and strong. It is, for
example, obvious in the building behaviour of invertebrates; the
caddis larva, *Lepidostoma hirtum*, which cuts more or less rectangular pieces of leaf that are convex at the front and sides and
concave at the back (see Fig. 1.10, Hansell 1972), is just one of
hundreds of examples in that insect order alone.

The use of hooks on the tools of these crows (Hunt 1996)
'suggests an appreciation of tool functionality'. However, in that
case, might not the rufous piha *appreciate* that vine tendrils will
lock together to make its nest platform or, since the hooks on the
Pandanus leaf tool are the spiny covering of its own surface, might
not the bushtit *appreciate* that the choice of a spiny lichen (see Fig.
4.11) will complement the spider cocoon silk to form a *velcro*
textile? Equally, the claim that the New Caledonian crow creates
two distinctly discrete tool types should be matched against the
choice of the long-tailed tit of four distinct and specialised building
materials (silk, moss, lichen and feathers) (Hansell, in preparation).

The importance attached to the crows' behaviour (Hunt 1996)
depends critically upon the definition of a tool, the currently accepted definition of which was given by Beck (1980). This definition is
not simple, but contains several elements. The tool must not be
part of the animal's body (a beak is not a tool); the user must
manipulate the tool in some way for it to realise its function; and,
finally, a tool cannot be attached to the substrate. This is a fairly
clear definition, but does seem to produce some rather arbitrary
distinctions (Hansell 1987b). The spider *Dinopis*, for example,
makes a small web which it holds in its legs, thrusting it down on
passing ants. This is a tool, but all other webs, however complex,

are not since they are anchored to the substrate. The woodpecker finch (*Camarhynchus pallidus*), that uses a fine stick held in the beak to extract insect prey from wood, is a tool user, but a shrike (*Lanius* sp.) that impales an insect on a thorn still attached to the bush is not.

It may still be argued that these distinctions, although fine, could serve to identify a particularly important aspect of animal behaviour, i.e. tool manufacture and use. The construction and even use of tools require manipulation. Also, the user needs to create a proper relationship between itself, the tool and the environment for the task to be accomplished. This gives three criteria for the special status of using tools: a high degree of manipulative skill, complexity in the organisation of the underlying behaviour and, finally, and probably the most striking attribute of tool use, the animal apparently perceives the object for something other than what it is and, simply by picking it up, transforms it (Hansell 1987b), e.g. the Egyptian vulture (*Neophron percnopterus*) picking up a stone to throw at an ostrich egg (van Lawick-Goodall & van Lawick 1966).

In fact, none of these three criteria confirms the special status of animal tools. The use of tools and even the construction of them generally provide the least impressive examples of manipulation for that group. The New Caledonian crow is probably the prime example of tool manufacture and use in birds (Hunt 1996); however, a scattering of other examples are known. The near relative of crows, the northern blue jay (*Cyanocitta cristata*), in captivity has been seen to tear pieces of paper which are then used to rake food within reach of its beak. The northwestern crow (*Corvus* (*brachyrhynchos*) *caurinus*) has been seen to use a stick like a woodpecker finch, as has the brown-headed nuthatch (*Sitta pusilla*) (Boswall 1983). African grey parrots (*Psittacus erithacus*) will hold a stick in one foot to scratch their heads (Boswall 1977), but all this is unremarkable in terms of manipulative skill compared to that shown by thousands of species of bird nest builders.

The argument of insight is equally weak. It is unnecessary to argue that the New Caledonian crow or Egyptian vulture *appreciate* the function of their tool. This was pointed out long ago (Alcock 1972). The fire ant (*Solenopsis invicta*) using a piece of dead grass as a sponge to mop up liquid food to convey to the nest (Barber *et al.* 1989) or the weaver ants *Polyrachis* and *Oecophylla* using larvae like tubes of glue to stick together leaves to make a nest (Hölldobler & Wilson 1990) all seem unlikely to be using insight, and there is no need to invoke any mechanism other than genetic to explain the occurrence of these behaviours (Hansell 1987b).

One example of bird tool use does certainly demonstrate learning by the tool user. That is bait fishing by the green heron (*Ardeola stricta*). Juvenile green herons throw a variety of objects onto the water surface, particularly leaves and twigs. These attract the attention of small fish, which the herons attempt to catch. More experienced, adult birds have apparently learned that flies are better bait at attracting fish and have also become more skilled at catching fish with any bait (Higuchi 1986). This is an interesting observation on foraging behaviour, but needs no explanation beyond acknowledging that the herons have improved their efficiency through associative learning, as do weaver birds in their nest building.

One cannot escape the conclusion that the fascination with tool making and using behaviour in all species, except the primates, is wilful anthropomorphism. It is significant that alongside the New Caledonian crow paper (Hunt 1996) is printed an assessment of its significance in the context of human evolution (Boesch 1996). The building attainments of mammals are poor, even making allowance for the beaver and some extensive burrowers. Tool manufacture and use in chimpanzees are well documented (McGrew 1992) and rightly raise important issues about their cognitive abilities, which Matsuzawa (1991), for one, believes are still underestimated.

Tool manufacture in higher primates has long been hypothesised to have been intimately connected in humans with the evolution of larger brain size and the emergence of complex languages (Washburn 1959, Parker & Gibson 1979), although more recent theories propose that increasing brain size was driven by the complex social demands of living in larger groups (Whiten & Byrne 1988). Whether or not tools were important in human evolution, there is no evidence that their influence on bird evolution has been other than marginal, certainly when compared to the influence of nest building. Tool construction and use are a footnote to the acknowledged manipulative skill and advanced nervous system of birds.

CHAPTER 5

The functional architecture of the nest

5.1 Introduction

The ground nesting blackstart (*Cercomela melanura*) is one of a number of species of mostly desert birds of several families that build a rampart of small stones at the entrance of the sheltered nest. Leader and Yom-Tov (1998), on the basis of artificial nest predation experiments, conclude this is to give incubating females early warning of the approach of nest predators and so save themselves. Stones are not a common nest building material, but they have the properties required for this defensive role.

Chapter 4 makes a case for the use of specific materials for constructional reasons, in particular allowing the behavioural repertoire to be kept simple, even stereotyped. However, uniformity in the choice of materials may also improve structural integrity. Uniformity of composition or standard building units should tend to eliminate points of weakness in the structure. This argument applies only to those parts of the nest that have an important structural role, but nests are differentiated structures; stones appear to be specifically chosen by the blackstart for a defensive role in the nest. So, there is also an argument for nest material specialisation on the grounds that the material chosen is the best available one for the job, structural or otherwise.

Nests perform a variety of functions under the general heading of protecting the eggs and young, different parts having different roles. In Section 3.3, four nest zones were identified. Nests located in trees, for example, have the engineering problem of containing the eggs; this must be solved with a *structural* nest layer and possibly also an *attachment* device to anchor the nest in position. Materials chosen for these parts of the nest should be able to withstand significant stresses of compression and tension. However, there are other aspects of the functional design of the nest which do not require such materials. These non-structural materials are located in parts of the nest where they alter the internal condition of the nest cup or cavity, or alter the external appearance of the nest. These are identified in Section 3.3 as materials of, respectively, the *lining* and *outer* nest layers. This chapter is an examination of the material composition and functional design of nests based substantially on data obtained from the nest survey, using the system described in Chapter 3. Each nest zone is examined, in the order outer nest layer, attachment, structural layer, lining.

The purpose of the examination of the data from the nest profiles is to obtain insights into functional design. This is done by relating aspects of the nest to one another, and relating nest features to the birds that build them, or to ecological variables such as the nest site. In recent years comparative studies have been criticised for the possible lack of independence of data points representing different species, on the grounds of their common ancestry; the solution advocated is the use of methods of analysis that take into account the phylogeny of the group being studied (Felsenstein 1985, Harvey & Purvis 1991). However, concerns have also been expressed that the phylogenetic methods may, at the same time as increasing the independence of points in the sample, introduce errors of their own, due to assumptions that may not be met or uncertainty over the phylogeny itself. Rosenzweig (1996), in a paper considering the possible association between coloniality and rate of speciation in birds, argues for a less dogmatic view of the use of phylogenetic methods, and Ricklefs and Starck (1996), in a paper comparing traditional comparative methods with phylogenetically based methods, conclude that the results of the two are often much the same.

There are circumstances in which the risks inherent in the more traditional comparative methods can be regarded as small. The sampling system itself can be conducted in a way that minimises the confounding effects of phylogeny, notably by sampling diversely across the group under study and at the level of genus rather than species (Felsenstein 1985). Both these principles were used in the bird nest survey. In addition, confounding effects of phylogeny will be weak if the traits under study respond rapidly to natural selection with little evidence of phylogenetic constraint. Evidence suggests that this may generally be the case in features of bird nests. Firstly, there appear to be few aspects of nests, in terms of either their design or composition, which are not to be found in a number of unrelated families (Section 5.4h), the use of arthropod silk in nests (see Table 4.1) and the 'hammock' nest design (see Table 5.2) providing good examples. Secondly, fitting nest designs to independently established phylogenies (see Figs. 9.3, 9.4 and 9.5) appears to indicate that changes of nest design can occur readily. For these reasons, the comparative analyses used in this chapter do not correct for phylogeny. More detailed analysis concentrates on three passerine families rather than the whole nest survey, as they provided reasonable sample sizes of related species; these were Tyrannidae, Corvidae and Fringillidae (Sibley & Monroe 1990).

5.2 The outer nest layer

In Section 3.3, the outer nest layer was defined as a layer with no significant structural role, applied to the outside of the structural layer. One effect of such a layer is to make the nest look different, implying that its function is to make the nest less obvious to visually hunting predators; however, protection from water penetration and temperature regulation are possible alternative explanations. The external appearance of the nest may also be altered in a much more radical way by the addition of large amounts of material which completely alter the overall shape of the nest. These materials may be added to the top of the nest well beyond the extent of a simple roof or overhead attachment, or may hang down well below the outer curvature of the nest floor. These are defined as *heads* and *tails* (see Fig. 3.1).

a) Nest decoration and nest size

For nests of arboreal species, concealment, by whatever device, is more likely to succeed in small rather than large nests. For larger species there may also be the possibility of deterring predators by active nest defence. For both these reasons, the expectation is that an outer nest layer will be more characteristic of small rather than large birds and their nests. This was tested on the nests of three passerine families: Tyrannidae $(n = 28)$, Corvidae $(n = 39)$ and Fringillidae $(n = 75)$. A logistic regression of the presence or absence of outer layer materials against nest weight, cup diameter and family showed no effect of family or nest weight, but a significant negative effect of nest cup diameter on the presence of outer layer materials (Humphries & Ruxton, unpublished analysis). The prediction that outer nest decoration is a feature of smaller nests therefore receives some support.

b) Frequency and type of decorative materials

Examination of a sample of 47 Tyrannidae, 65 Corvidae and 91 Fringillidae for materials in non-structural or outer, decorative layer showed that the great majority of nests possessed no such layer: respectively 68%, 84% and 80% (Fig. 5.1). The external appearance of the majority of nests is therefore determined by

Figure 5.1
Number of categories of
materials identified from
the four nest zones:
(a) outer (decorative)
layer, (b) attachment,
(c) structural layer and
(d) lining. ('missing'
denotes absence of
attachment evidence in
museum specimen.)

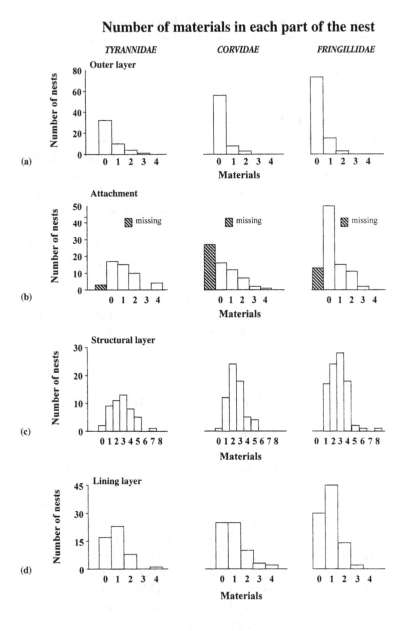

structural layer materials, although their selection could in part be determined by nest detectability. Oniki (1985), in a survey of nests in an Amazonian site, found that nests of pale materials were associated with open sites and dark nests with darker, shaded sites.

Of the 45 nests with decorative materials, 33 (73%) had only one material, and in 41% of the externally decorated nests, lichen flakes were one of or the sole material. The second most commonly

Figure 5.2
The nest of the *blue-grey
gnatcatcher (*Polioptila
caerulea*) is covered
externally by a more or
less continuous layer of
flakes of pale lichen.
(Photograph by M.
Hansell.)

used material (30%) was moss, and the third (27%) was either
pieces of white membranous silk or white spider cocoons.

The lichens typically have a pale greyish-green appearance and
may be applied to the nest surface in a more or less continuous
layer, as in the nest of the *blue-grey gnatcatcher (*Polioptila
caerulea*) (Fig. 5.2). In the *long-tailed tit (*Aegithalos caudatus*) the
layer is composed of over 2000 lichen flakes interspersed with 150
small pieces of white membranous spider cocoon ($n = 17$) (Hansell,
in preparation). The nest of this species is described by Perrins
(1979) as being 'beautifully camouflaged, especially against a tree
trunk'. Collias and Collias (1984) describe the outer lichen layer on
nests as causing them to 'resemble the immediate environment',
citing the nest of the *blue-grey gnatcatcher as an example. They
also state that 'often, lichen covered nests of various birds are
saddled on a branch and look like a knot on that branch'.

Where the predator can detect, in this case, a nest as an object
but perceives it as part of the tree, the concealment mechanism is
regarded as a type of *mimicry* by Robinson (1981), who reserves the
term *crypsis* for objects that blend into the background. Endler
(1981) refers to resemblance to part of a plant as *masquerade*, using
crypsis for resemblance to 'a random sample of the background'.
Confusingly, Starrett (1993) calls both types of concealment crypsis, while agreeing that they may be perceived differently. The
important point here is that the *branch matching* explanation of
Collias and Collias (1984) is distinct from a general background
matching hypothesis, and is referred to here as a masquerade
hypothesis.

For nests positioned in trees, as virtually all lichen-covered nests are, the background into which the nest would merge if crypsis were the concealment mechanism may not be other vegetation but the sky. The hypothesis of background matching to the sky or to a light background beyond the nest site does fit the evidence that lichen coverings give the nest a generally pale appearance. More significantly, it does help to explain the mixing of white membranous spider cocoons with lichen, and even the otherwise bizarre, completely white outer coat of membranous spider cocoons on the nests of, for example, the *solitary vireo (*Vireo solitarius*) (Harrison 1979) and of the *crimson-backed sunbird (*Nectarinia minima*) (Hansell 1996b).

For the branch matching (masquerade) hypothesis to be upheld, lichens on the nest should be associated with lichens on the nest attachment site; materials other than lichen applied to the nest exterior should tend to match the appearance of the attachment site; and nests coated with lichens should be attached to branches of similar or larger diameter than themselves. For the light reflection hypotheses (crypsis hypothesis) to be upheld, lichen on the nest may occur in the absence of any lichen on the branch to which the nest is attached; other materials on the nest surface should enhance light reflection; and nests may be attached to branches much narrower than the nest diameter and so bear no resemblance to part of the branch.

Hansell (1996b) tested these hypotheses using evidence obtained from a sample of *long-tailed tit and *blue-grey gnatcatcher nests, and from examination of single nests of over 50 other species identified in the nest survey. Only 16% of *long-tailed tit nests were on branches bearing lichen; however, all the nests bore lichen on the surface, indicating the absence of an association between lichen on branches and on nests. *Blue-grey gnatcatcher nests varied in the amount of lichen on their surface and there was also variation in the amount of lichen on the branch where the nest was attached; however, there was no evidence of an association between the two, and half of the *blue-grey gnatcatcher nests were on branches that bore no lichen.

Twenty-eight species belonging to eight families were found to apply lichen to the exterior of the nest, and 19 belonging to eight families to include white membranous spider cocoon silk (nonstructural) either with lichen or alone (Table 5.1, Hansell 1996b). The nest of the *plain-throated sunbird (*Anthreptes malacensis*) had an almost complete covering of pale silver-grey cuticle of skeletonised leaf, apparently similar to pieces of 'silvery' bark

Table 5.1. The families and number of species within them of:
(a) nests bearing external application of lichen flakes, and
(b) nests bearing external patches of white, membranous spider
cocoon silk

Family	Species
(a) Nest exterior with lichen flakes	
Trochilidae	15
Tyrannidae	2
Vireonidae	2
Corvidae	4
Bombycillidae	1
Certhiidae	1
Aegithalidae	2
Fringillidae	1
(b) Nest exterior with white patches of spider cocoon silk	
Tyrannidae	3
Meliphagidae	1
Eopsatriidae	2
Vireonidae	3
Corvidae	4
Aegithalidae	2
Nectariniidae	1
Fringillidae	3

illustrated by Campbell (1901) on the exterior of the nest of lemon-bellied flyrobin (*Microeca flavigaster*) Eopsaltriidae. Of the *long-tailed tit nests examined, 45% included pieces of white paper or fragments of white polystyrene foam in the outer coating, one nest bearing 1568 polystyrene spheres, whereas only one specimen bore neutral materials. Pieces of near-white paper were found on the surface of 19% of the 32 nests of the *bushtit (*Psaltriparus minimus*), a feature also recorded by Addicott (1938). White paper fragments were also found on the exterior of the nests of three species of vireo, and white or pale blue fragments of paint were found applied to the exterior of two nests of *Anna's hummingbird (*Calypte anna*) (Hansell 1996b).

The diameter of the nests of 25 species bearing lichen on their outer surface was in all cases considerably greater than that of the thickest branch supporting them (Fig. 5.3). The majority of lichen-bearing nests therefore bear no resemblance to a knot on the branch supporting them, thus complying with the light reflection hypothesis (crypsis) rather than with the branch matching hypothesis

Figure 5.3
The diameter of the 25 lichen-bearing nests, each of a different species, plotted against the diameter of the branch on which each was supported. All nests represented by open symbols were supported by branches that bore no lichen. Solid symbols represent nests attached to branches bearing lichen. A 'line of equality' shows the line upon which points would lie if branches had the same diameter as the nests that they supported. (Adapted from Hansell 1996b.)

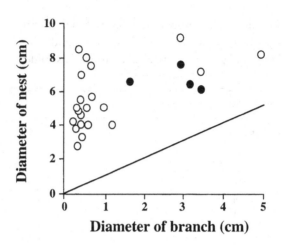

(masquerade). The merit of this concealment mechanism seems to be that the nest, which would otherwise appear to a predator as a dark mass, tends to dissolve into the paler background. This does not mean that masquerade cannot explain the concealment mechanism of the nests of some species. The nest of the *varied sittella (*Daphoenositta chrysoptera*) is made of silvery flakes of bark and is sited in the fork of a tree of the same bark (Campbell 1901). The nest of the *long-tailed tit, when supported in a bush by several small branches, does not look like part of the bush; however, it has an alternative site, high in the forks of silver birch branches, where the pale colour of the lichen does blend with that of the tree. Empirical evidence from baited model nests in the field is now needed. Alternative but untested hypotheses should not, however, be dismissed; reflection of radiation as a temperature control mechanism is a possible explanation for the same device. Equally, an outer covering of lichen could prevent rain penetrating the nest. These hypotheses are not mutually exclusive, so more than one of them may be supported by further research.

Lichen and white spider cocoons may be applied to nests, not as a continuous mosaic covering but as many or a few conspicuously pale spots against the darker background of the structural layer of the nest. This is such a prominent feature of the nests of some species, for example those of the paradise-flycatchers (*Terpsiphone*) and of the *cocos flycatcher (*Nesotriccus ridgwayi*, Tyrannidae), that it invites a different kind of concealment explanation. The hypothesis of Hansell (1996b) proposed that the cocoons or lichen flakes are light-reflecting spots which, on the surface of a darker shape, create the impression of light passing through an insubstantial object unworthy of a predator's attention. The same explanation is given by Cott (1940) for the white spots on the body of the forest-dwelling bush buck (*Tragelaphus scriptus*) and included under the heading *disruptive camouflage*, since they have the effect of breaking up the solid shape of the object which is seeking concealment. This explanation is also given by Oniki (1985) for the colour of eggs in bird nests: white eggs associated with thin nests through which the eggs are visible from below yet resemble the holes themselves, and blue eggs in thick, dark cup nests which resemble sunlight spots on dark foliage. It seems, therefore, that the application of lichen flakes, white spider cocoons and occasional other material to the outer surface of the nest may act to conceal the nest through crypsis, through masquerade, through disruptive camouflage, and possibly combinations of all three. However, empirical evidence is lacking and completely different explanations also deserve investigation.

In the case of open nests, a predator will see not simply the nest but the bird sitting on the nest, so both should be considered together when judging the nature of camouflage. This has never been done in any systematic way, but it is conspicuous that the head and neck of the pied monarch (*Arses (telescophthalmus) kaupi*), for example, are prominently patterned in black and white and the bird has a blue eye ring; all these are clearly visible when the adult or feathered chick sits in the diminutive nest (Fig. 5.4). However, not only is the shape of the bird broken up by the contrast of its black and white markings, but the nest shape also is broken up by the application of lichen or spider cocoons. It seems that the bird and nest together have an integrated disruptive camouflage. On the other hand, Martin (1993c) points out that ground-nesting species such as the red-faced warbler (*Cardellina rubrifrons*) and orange-crowned warbler (*Vermivora celata*), whose open nests are made of materials which generally match the substrate, are themselves coloured in ways that allow them to blend in, the bird and the nest in this case together exhibiting concealment by crypsis. Haskell (1996), using artificial nests either bright red or dull brown in colour and baited with a quail egg, found that predation was greater in red than in brown ground nests, whereas no difference was observed between the two coloured nests placed in sites 1.5–2.0 m up trees, suggesting that different concealment rules may operate in the two sites.

c) Snake skin

The function of snake skin in the nests of some species is unknown and its location in the nest varies between species. The number of species in which snake skins are recorded is small, but they belong

Figure 5.4
Disruptive camouflage:
the black and white
coloration of a pied
monarch chick (*Arses*
(*telescopthalmus*) *kaupi*)
and the white patches on
the nest all appear to
contribute to the breaking
up of their outline.
(Photograph courtesy of
Oxford Scientific Films.)

to at least five families. Among them are species of cavity nesting Tyrannidae, for example the western kingbird (*Tyrannus verticalis*) and *Myiarchus* species (Harrison 1978), and of Trogloditidae, for example the house wren (*Troglodytes* (*brunneicollis*) *aedon*) (Rowley 1966), and Bewick's wren (*Thyromanes bewickii*) (Harrison 1978). Skutch (1960) attaches no significance to the collection of these materials other than that they are 'soft and pliable', because Cellophane or plastic may be selected as well as or instead of snake skin. However, an alternative hypothesis is that snake skin acts as a deterrent to nest predation by small mammals that might become victims of snakes. This allows snake skin to be considered here as a nest decoration.

Some evidence does support the predator deterrent hypothesis for the snake skin. On the cup nest of the *blue grosbeak (*Guiraca caerulea*, Fringillidae) and the *paradise riflebird (*Ptiloris paradiseus*, Corvidae) the snake skin is placed on the outside of the nest (respectively, Harrison 1979, North 1901–4). Also, Scott *et al.* (1977) note that, while snake skin may be placed in the nest cavity by the great crested flycatcher (*Myiarchus crinitus*), it may also be placed 'dangling from the cavity opening'. ffrench (1991) records that the *pale-breasted spinetail (*Synallaxis albescens*, Furnaiidae), which makes a domed twig nest with narrow horizontal entrance passage, places the snake skin 'in the entrance passage'.

d) Heads and tails

The nest of the *common waxbill (*Estrilda astrild*) may bear a smaller domed chamber above the egg chamber, interpreted as a bedchamber for the male bird (Sick 1993). The nest of the yellow-rumped thornbill (*Acanthiza chrysorrhoa*) has a cup on top of a domed nest chamber, suggested by North (1901–4) to be a roosting place for the male. Both these devices might equally act as false nests to distract predators that have detected the nest, an interpretation that is generally accepted for the blind, false entrance just below the real entrance to the domed nest of the Cape penduline tit (*Anthoscopus minutus*) (Collias & Collias 1984).

The extension of the outline of the nest by the addition of material to the roof (head) or to the base (tail) provides additional examples of what could be nest concealment devices (Fig. 5.5). Architecturally as well as taxonomically, these are a miscellaneous collection so, in the absence of any field data, speculation can be justified only as of heuristic value. The dangers of functional speculation are especially obvious when the oddities of the nest are small; the nest of the *common tody-flycatcher (*Todirosturm cinereum*), for example, has a tuft of material extending below it (Skutch 1960). Another Tyrannid, the *scale-crested pygmy-tyrant (*Lophotriccus pileatus*), has a slightly more obvious extension to the top and more materials added below. However, the extension to the top could simply be described as the suspension and that to the bottom simply as a byproduct of the method of construction and of no functional significance. More extreme designs do, however, invite separate explanation. For example, the nest of the *red-faced spinetail (*Cranioleuca erythrops*, Furnariidae) has, long extensions above and below, which are solid masses quite separate from the more or less spherical wall of the nest. In two further Tyrannidae, the *eye-ringed flatbill (*Rhynchocyclus brevirostris*) and *black-tailed flycatcher (*Myiobius atricaudus*), the long cone of material below which the nest is suspended contains fine woody stems, rootlets, sticks, and other materials.

The nest of the garnet-throated hummingbird (*Lamprolaima rhami*) is a rather massive cup, 6 cm across, perched on a root under a bank, but has hanging below it a few strands of vegetation (Rowley 1966). This could be mistaken for a lump of earth with a few rootlets hanging from it. This could also explain the attachment

Figure 5.5
'Heads' and 'tails' appear
to serve as devices that
distort or break up a
typical nest shape. Tail
examples:
(a) rufous-backed fantail
(*Rhipidura rufidorsata*);
(b) *scale-crested
pygmy-tyrant
(*Lophotriccus pileatus*);
(c) *long-tailed hermit
(*Phaethornis
superciliosus*). Head and
tail example:
(d) *red-faced spinetail
(*Cranioleuca erythrops*).
Head examples:
(e) *eye-ringed flatbill
(*Rhynchocyclus
brevirostris*);
(f) *black-tailed
flycatcher (*Myiobius
atricaudus*).

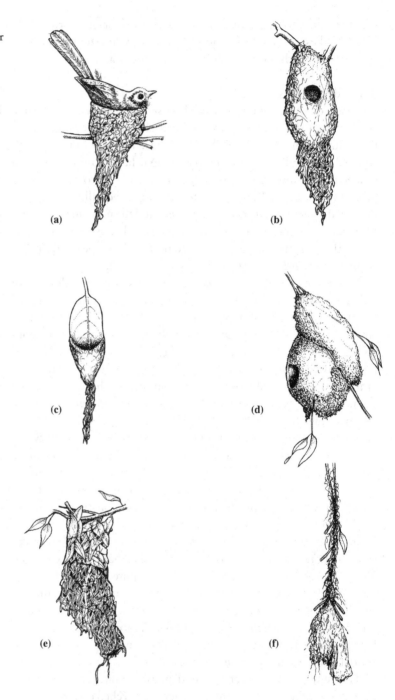

(a)

(b)

(c)

(d)

(e)

(f)

of the nest of one year to that of the previous year to create a less regular overall shape (Rowley 1966), a behaviour also shown by some other hummingbirds, e.g. the *grey-breasted sabrewing (*Campylopterus largipennis*). These hummingbird nest examples therefore suggest the interpretation of concealment by masquerade, although they are possibly just examples of improved nest attachment. Skutch (1960) reports the nest site of the Central American wren (*Troglodytes ochraceus*) as located inside a tangle of irregular vegetation resulting from a branch rotting through and a bundle of epiphytic growth left hanging by one or two vine stems. In this case the nest does not have a constructed head or tail; however, other species, such as the Tyrannidae already cited, may build their nests to mimic this phenomenon, as does the rufous-sided broadbill (*Smithornis rufolateralis*, Eurylaimidae) in Africa (Chapin 1953) and the rufous babbler (*Pomatostomus isidorei*, Pomatostomidae) in New Guinea (Coates 1990), both of which make nests with irregular hanging tails. Another location for irregular hanging masses of material in tropical environments is over streams or rivers, where periodic spates leave streamers of flotsam on the ends of branches overhanging the water. In such a site is found the nest of the royal flycatcher (*Onychorynchus* (*mexicanus*) *coronatus*), 'a yard or two in length...' and 'one of the most amazing structures made by any of the... Tyrannidae...' (Skutch 1960). Masquerade therefore does appear to explain the presence of heads and tails on a variety of nests that are often quite large in size and, while difficult to conceal from predators, can be made to resemble naturally occurring objects.

There do, nevertheless, remain a miscellany of additional nest structures that may require other sorts of interpretation. The nest of the yellow robin (*Eopsaltria australis*, Eopsaltriidae), for example, is a bulky cup from which hang a ring of long pennants of bark attached to the side by silk (North 1901–4); the white-browed robin (*Poecilodryas suerciliosa*), of the same family, shows the almost identical feature (Simpson & Day 1989). Here, the structure looks so regular as to attract attention, but perhaps this chandelier design looks different when the bark strips swing in the wind.

The tail below the hanging nest of the sooty-capped hermit (*Phaethornis* (*pretrei*) *augusti*) is open to a different interpretation. The nest cup is attached to the silk suspension at one point on the nest lip (Fig. 5.6, Skutch 1973), and would therefore tip over were it not for the numerous lumps of dry clay and pebble suspended below the nest as a counterweight. A similar principle could be involved in stabilising nests sighted on top of a branch and therefore

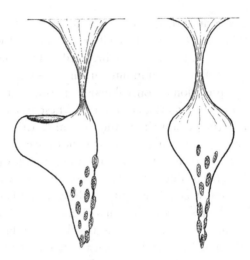

Figure 5.6
The 'tail' hanging below the nest of the sooty-capped hermit (*Phaethornis (pretrei) augusti*) is weighted with mud pellets, apparently as a counterweight to keep the cup level when occupied with parent or brood. (Adapted from Skutch 1973.)

in danger of toppling over. An example of this could be the tail projecting below the nest of the rufous fantail (*Rhipidura rufifrons*), described by North (1901–4) as 'like a wine glass with the base broken off at the lower end of the stem' (see Fig. 5.5).

5.3 Nest attachment

Nest attachment devices are solutions to problems of anchoring a nest in the chosen nest site. For nests on the ground this is clearly not a problem; for nests supported from above (*top, top lip*), or at least not supported from below (*top side, bottom side, wall* and *bottom multiple vertical*), they are essential; whereas for nests above ground but supported as *bottom* or *bottom multiple branched* they are not essential but may be present.

In the nest survey there was a strong possibility of sampling bias, because large nests are less likely to be collected by museums, and the majority of nests are collected without evidence of the nest site. For a number of the nests examined it was impossible to decide whether attachment materials had been present, possibly leading to an underestimate of the occurrence of certain types of attachment. Nevertheless, the specimens examined do allow some assessment to be made of the nature of attachment materials and how they are used in the different classes of nest site.

a) Number and type of materials in attachments
In the sample of nests of Tyrannidae, Corvidae and Fringillidae, a proportion of each had to be omitted from the analysis of attachment types due to the absence of clear evidence of whether an

attachment had or had not been present, leaving a sample of 46 Tyrannidae, 38 Corvidae and 78 Fringillidae with or without attachments. These showed that:

1.   About half of the nests in the three families overall have no attachment materials (Tyrannidae 37%, Corvidae 42%, Fringillidae 64%).
2.   Attachments when present are simple in composition, 89% with one or two categories of material only (see Fig. 5.1).
3.   In the great majority of nests with attachments (85% within these three families), attachment materials are the same as or a subset of structural layer materials.
4.   A wide variety of materials can be used in attachments. Of the 33 categories of nest material, 18 (55%) occur at least once as attachment materials.
5.   Of the materials scored in attachments, 46% were one of the categories of arthropod silk.
6.   In a separate analysis (Section 5.4e), silk as an attachment material was shown to be characteristic of lighter nests and birds.

b) Attachment type and nest support diameter

An interpretation of the attachment of nests so that they hang below a branch (top and top lip), rather than the support being placed under the nest, is that top-supported nests are responding to predator pressure by being located on finer branches, towards the edges of trees. From this two predictions can be made. Firstly, top-supported nests will tend to be small and, therefore, all other things being equal, smaller and lighter than bottom-supported nests. Secondly, the diameter of the branch supporting hanging nests will be less than that of the branch supporting nests from below.

No significant relationship was found between nest cup diameter or nest weight and support diameter for the three families Tyrannidae, Corvidae and Fringillidae. Nor could any effect of nest support type (above or below) on support diameter be demonstrated (Humphries *et al.*, in preparation); however, sample sizes were small ($n = 17$ and 16 respectively for the total of the three families). This emphasises the need for collecting nests attached to their nest sites or systematically measuring nest support diameters in the field.

c) Occurrence of attachment types

Top lip attachment. Twenty-six nests fastened in this way were found in the total survey of 518 nests, representing three slightly different designs. Dominant among these (23 species) is the hammock, fastened by the rim in the angle of a forked twig; the second is essentially the same, except that the nest cavity is deepened to form a pouch with an entrance at the top. The third variant is a deep pouch, suspended at the lip by a single, rather narrow attachment (e.g. see the oropendola nest in Fig. 4.15). Silk was used as one of the attachment materials, or the only one, in 18 species (75%). Apart from silk, only one category of materials was used for attachment in more than five of these nests; this was horsehair fungi, found in the suspensions of 11 of the nests and, in five of these, not in association with silk. The remainder of the non-silk materials generally had a fibrous quality (palm fibre, grass leaf, rootlets) except for one nest supported by coiled vine tendrils (*rufous piha, *Lipaugus unirufus*, Tyrannidae, see Fig. 4.7).

The nest survey recorded the hammock nest attachment in eight passerine families and it is described in others (Table 5.2), in some of which, like the Vireonidae and Thamnophilidae, it is apparently the typical design. The degree of standardisation in the choice of site in a horizontal, forked twig could allow a rather highly specialised attachment design to evolve. This is suggested by the observation of a *four attachment point* design (Fig. 5.7) seen in five species from three different families: the *collared antshrike (*Sakesphorus bernardi*, Thamnophilidae), the *silver-throated tanager (*Tangara icterocephala*, Fringillidae), the *lance-tailed manakin (*Chiroxiphia lanceolata*, Tyrannidae), the *red-capped manakin (*Pipra mentalis*, Tyrannidae), and the *bright-rumped attila (*Attila spadiceus*, Tyrannidae).

Top and top-side attachment. Nests attached in these two ways represent the varied possible solutions to nests suspended wholly or predominantly from above. Thirty nests out of a total nest survey were designated as attached from the top, all roofed, with or without an entrance tube, and representing nine passerine families (Sibley & Monroe 1990). They were suspended by materials belonging to 16 of the 33 material categories, and in 11 of the 30 nests (37%) silk was one of the suspension materials. The versatility of silk as an attachment material is particularly evident in the specialised attachment of the domed nest of the *origma (rock warbler) (*Origma solitaria*) directly to the roof of a cave (North 1901–04) by means of silk bound round small, rocky projections (Fig. 5.8).

Nine nests were designated as having a top-side attachment, seven of which were hummingbirds' nests attached to the sides of drooping leaves. These are not wall-attached nests because the

Table 5.2. **Eight families of birds in the nest survey in which at least one species built a 'hammock' or deeper pouch-hammock design suspended between the arms of a forked twig by means of a top-lip attachment**

Family	Number of species
Thamnophilidae	4
Tyrannidae	8
Meliphagidae	1
Vireonidae	3
Corvidae	1
Muscicapidae	1
Zosteropidae	1
Fringillidae	4

Note. At least one other family uses this design, which is recorded for the icterine greenbul (*Phyllastrephus icterinus*, Pycnonotidae) (Chapin 1953).

Figure 5.7
Detail of one attachment in a four attachment point design shown by the hammock nest of the *collared antshrike (*Sakesphorus bernardi*, Thamnophilidae). The drawing below shows the overall nest design and the area covered by the photograph. (Photograph by M. Hansell.)

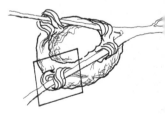

Figure 5.8
Detail of part of the silk
attachment of the domed
nest of the *origma (rock
warbler) (*Origma
solitaria*) looped over
mineral nodules on a cave
roof. The drawing below
shows how the nest is
suspended directly from
the roof of a cave.
(Photograph courtesy of
Glasgow University
Photographic Unit.) (Line
drawing by M. Hansell.)

attachment material is bound round the nest and round the back of
the leaf. In five of these hummingbird nests, the attachment
materials were wholly silk, although one included no silk. The
remaining two nests are of the *African palm swift (*Cypsiurus
parvus*) and the *sepia-capped flycatcher (*Leptopogon amauro-
cephalus*, Tyrannidae), attached to a hanging vine stem.

Bottom attachment. Nests attached in this manner sit on top of a
single horizontal branch or on top of a horizontal branch at the
point where it forks. The attachment problem is to prevent the nest
from toppling off the support provided below it. Eleven nests in the
total nest survey were given this designation, three belonging to the
non-passerine family Trochilidae (the hummingbirds), the other
eight to three families (Sibley & Monroe 1990) of passerines,
Tyrannidae, Corvidae and Fringillidae. In all of these specimens,
attachment was by the use of silk alone. This appears to be effective
for a combination of reasons, not all of which are shown in all
cases. The nests are quite small (range 3.7–42.0 g), and the support
branches are quite thick, with the bottom of the nest moulded to the

branch upper surface. The nests of some hummingbirds which are sited in this way are attached by numerous silken guys from the cup wall to the surface of the branch (see Fig. 3.6, Rowley 1966).

Bottom multiple (vertical). A particular attachment problem is faced by birds which place their nests in stands of reeds or giant grasses, because the attachment sites are vertical and only a few millimetres in diameter. The solution to this problem is to bind several of these supports into the wall of the nest, providing sufficient grip to prevent the nest slipping down. In the total nest survey, nine examples from five families (Furnariidae, Paridae, Passeridae, Sylviidae, Fringillidae) were assigned to this category, four cup shaped and five domed. In only one of these, the *Eurasian reed-warbler (*Acrocephalus scirpaceus*, Sylvidae), was silk used, the remainder using a variety of vegetation attachment materials only. The nest of the *scarlet-headed blackbird (*Amblyramphus holo-sericeus*, Fringillidae) is apparently made by enveloping the reed stems in wet dead leaves which, on drying, secure the substantial cup nest in position. The nest of the *wren-like rushbird (*Phleo-cryptes melanops*, Furnariidae) is a hard, domed shell of broad grass leaves and aquatic vegetation embedded in mud; a 75 g nest is the heaviest in a category of nests ranging from 11 g to 75 g.

Wall. Nests of this type are attached to a rigid and solid vertical surface. This may be a rock surface or inside a tree cavity. Only nine species examined were in this category. They belonged to two families, the non-passerine swifts (Apodidae) and the swallows and martins (Hirundinidae). All these nests are secured using an adhesive, i.e. a plastic material that hardens, either alone or in combination with a fibrous material, the former mucus, the latter mud. Mud is known as an adhesive attachment material for cliff-attached or wall-attached nests of three passerine families Tyrannidae, Hirundinidae and Picathartidae, respective examples being the *eastern phoebe (*Sayornis phoebe*), *cliff swallow (*Hirundo pyrrhonota*), and the substantial nest of the white-necked rockfowl (*Picathartes gymnocephalus*) (Grimes & Darku 1968, Section 6.5). The principle of adhesion in these nests is likely to be the simple mechanical process referred to as *hook and eye* (Howard 1997), whereby the mud or mucus flows into the irregularities and crevices of the attachment surface, locking the nest in position as it hardens.

Bottom multiple (branched). This was by far and away the commonest category of attachment, with 33% of all surveyed passerines included in this group. It was also the category which, using museum specimens, produced the greatest problems in assessing the nature of any attachment materials when supporting branches were not provided. Nests of 54 passerine species in the survey had identifiable attachment materials. A further 70 were classified as supported in this way from museum collection notes, but without the supporting branches to confirm whether attachment materials had been present. Nests with identified attachment materials were drawn from 15 passerine families, and were predominantly cup shaped (85.2%). Fifteen out of the 33 possible different categories of attachment materials were recorded, including, silk, hair and mud as well as a variety of plant materials, although silk was again the most common attachment material, used in 83.3% of nests.

5.4 The structural nest layer

Nests must not fall apart during their period of use This necessitates all nests, except some cavity or ground nests, having a structural nest layer to retain the nest shape and integrity. The forces acting on the nest will vary due to the support provided by the nest site and the forces exerted by the occupants inside it.

The nest support can be regarded as predominantly from below (ground, bottom multiple) or above (top, top lip) or from the side (wall, top side, bottom side). The mud cup of the nest of the *white-winged chough (*Corcorax melanorhamphos*) is a *bottom-supported* cup with vertical walls. Mud is strong in compression so it can stand its own weight acting vertically downward and, since grass and strips of bark are embedded in the wall, it can withstand tension in the walls resulting from radial forces, produced by an incubating bird or chicks within the nest cup. A bottom-multiple-supported stick nest such as that of *Florida jay (*Aphelocoma coerulescens*) may need relatively less strength to oppose these outward forces as the multiple-forked branch in which the nest rests resists them. Platform nests made of twigs, like that of the *wood-pigeon (*Columba palumbus*), are made of an array of horizontal sticks acting as beams on which the birds stand. Here, the short distance between points of support along any beam means that the bending forces acting on it at any point are likely to be low (Gordon 1978).

A side-attached nest is a cantilevered beam with the upper rim and point of attachment in tension and the lower surface and point of attachment in compression. In the pure saliva nest of the *edible-nest swiftlet (*Collocalia fuciphaga*), the one nest material resists both forces. For the top-side attached nest of the *long-tailed hermit hummingbird (*Phaethornis superciliosus*) it is the plant down which forms the bulk of the cup; this resists compression, while the enveloping swathe of silk that secures it to the dangling leaf resists the tension. Hanging dome nests supported at the top or hammock nests with top-lip attachment are nests predominantly in tension, resisting stresses from the weight of the nest and its contents which find no support below them.

a) Nest weight and nest design in relation to bird weight

Larger birds, particularly when building nests in trees, may be expected to build nests to bear greater stresses from parents, eggs and chicks than smaller species. If these mechanical considerations are important in nest design, then there should be a positive correlation between bird weight and nest weight. This was examined in a sample of 31 Tyrannidae, 42 Corvidae and 76 Fringillidae for which nest weights were available from the nest survey and bird weights from Dunning (1993). Analysis of covariance confirms the significant positive relationship between nest weight and bird weight (Fig. 5.9), and between-family differences, even correcting for the fact that birds of these families may differ in weight. A similar general relationship was found between nest size measurements and bird weight or wing length in a sample of European passerines (Slagsvold 1989b). The Tyrannidae appear to have more species above the line for the total population than below, indicating a trend (albeit weak) to build heavier nests than would be expected for their size. The Fringillidae, on the other hand, appear to build lighter nests than would be expected from the total population. The smaller species of Corvidae build rather lighter nests than expected, while those of the largest species are heavier.

The between-family differences could, in part, be influenced by differences in frequencies of domed and open nests in each of them. One-third of the Tyrannidae and 12% of the Fringilladae in the sample had domed nests, whereas all of the Corvidae had open cup nests. Excluding the Corvidae from the analysis, and comparing the weights of cup and domed nests, with bird weight kept constant, showed that domed nests do tend to be heavier than cup nests. This could explain the relatively large nests of the Tyrannidae. A logistic regression of dome/cup against bird weight and family showed no detectable effect of family, but that small birds showed a significantly greater tendency to build domed rather than cup nests (Humphries & Ruxton, unpublished analysis).

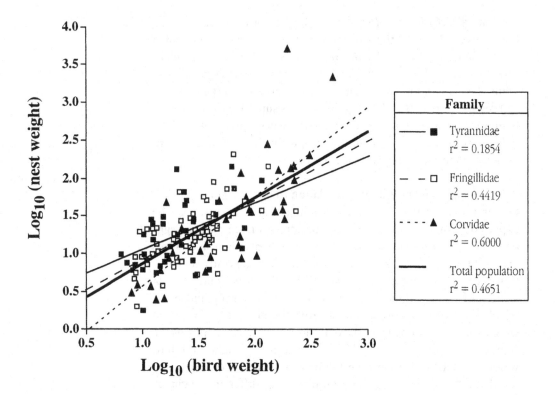

Figure 5.9
The relationship between weight of nests and their builders, for nests of Tyrannidae, Corvidae and Fringillidae. The slopes are shown for each of the three families, and for the three together (total population). (Humphries & Ruxton, unpublished analysis.)

In the Corvidae, the relatively heavy nests of larger species compared with smaller ones may reflect the effects of predation. Evidence of a positive correlation between nest size and predation is given in Section 5.5, implying that this will select for smaller nests, However, the study referred to was carried out on nests of relatively small species, for which size reduction may benefit concealment. For larger birds, on the other hand, concealment is a less realistic option; here, selection may favour more massive nests for stability, insulation or protection from predators.

b) Number of materials in the structural layer
Examination of the structural layer of the nests of 49 Tyrannidae, 65 Corvidae and 91 Fringillidae showed that, out of 33 possible categories of material, nests in the three families showed modal values of, respectively, three, two and three (see Fig. 5.1). Given that the modal value in all three families for the outer layer is zero and for the lining layer is one, nests prove to be generally of rather simple composition.

c) Associations of materials

The use of the *velcro* principle as a means by which silk binds together plant materials (Section 4.7c) suggests that certain plant materials might provide a particularly effective *hooked* material to bond with the *looped* threads of silk. If this were the case, then those materials might predominate in nests where there was a strong dependence upon silk. Equally, it might be argued that, if birds are standardising the selection of any materials for constructional or structural reasons (Section 4.9), then each material might have certain others with which it makes a particularly effective combination. In this case, an association between materials could be detected from the examination of the nest composition data as a clustering of materials in particular associations.

A simple qualitative assessment of associations was made by selecting four of the five most frequently occurring materials in all nests sampled (silk, grass stems, narrow grass leaf, and sticks), and scoring the occurrence of other materials with them. In the Fringillidae it was found that at least one of the three categories of silk occurred with 14 of the other 29 materials; grass stems, grass leaf and fine woody stems predominated among these materials, but moss, as used in the nest of the *long-tailed tit (see Fig. 4.10), was no more frequently associated with silk than were seven other materials. Narrow grass leaf occurred with 21 other categories of material, and grass stems with 19, although both were associated most strongly with one another, and then with fine woody stems. The number of stick nests in the sample was small (ten), but even here they were found to be associated with ten other materials, with fine woody stems predominating. A similar pattern was apparent for the Tyrannidae and Corvidae.

This level of analysis, therefore, does not reveal discrete groupings of nest materials. This does not necessarily mean that the majority of nests are unspecialised structures, although this may be so, but rather that more detailed examination of nests is needed to understand how they work. This is emphasised by nests of simpler composition, where it is evident that some birds are taking advantage of the properties of particular materials to produce elegant design solutions.

d) Standard units

In Section 4.9 it was pointed out that the repertoire of construction behaviour could be made simple and structural properties of the nest more consistent if the building materials used were uniform. If these constructional and structural considerations have exerted a significant selection pressure, then two predictions can be made. The first is that the structural layer of a nest should be made of few materials, some of which may have similar properties; the second is that some nests will be made of standardised building units or *bricks*, which render the construction process more simple. The first of these is generally born out by Section 5.4b, although qualified by Section 5.4c. The second can be seen clearly in at least a minority of nests.

Plants are made up of standard elements such as leaves and flowers; some of these have become a major component in the nests of quite unrelated species. An example of this is leaf petioles. Some species of trees have pinnate leaves which, when the leaves fall, shed pinnae from the petiole, which is then left as a tapering, somewhat flexible rod. The nest of the *greenish elaenia (*Myiopagis viridicata*, Tyrannidae) is built simply of leaf petioles held together with silk – two materials to construct the whole nest. The nest survey reveals a total of eight species from six families using leaf petioles as structural materials, and a further four species from three families using them as a lining material (Table 5.3).

Fern scales, the papery, pennant-shaped coverings of fern stems (see Fig. 3.8), were recorded as structural materials in four out of 30 (13%) species of hummingbirds' nests examined. Pine needles, a more obvious standard building unit, however, were recorded in the structural layer of only one species, the Indian swiftlet *Collocalia unicolor*.

Forty-eight species of birds from 13 families were found to be using plant down as a structural material to create a very simple nest. Fourteen of these (29%) were using plant down and some form of silk as the only other material, 27 (56%) were using silk in combination with at least one other type of plant material, while only 7 (14.6%) were using plant down with another plant material and no silk. The combination of a cup of uniform texture made entirely of plant down bound with silk was particularly evident in hummingbirds, with 11 of the 30 species examined (37%) having this composition. Standardisation through selection from an available range of materials is also widely evident from the examination of other structural materials such as silk, grasses and sticks, as can be seen in the succeeding sections.

e) Silk
Identity of silk in nests. One hundred and thirty-two silk samples from the nests of 110 species in six families (Trochilidae, Tyrannidae, Vireonidae, Corvidae, Sylviidae, Fringillidae) were examined

Table 5.3. **The use of leaf petioles as standard building units in either the structural or lining layer was recorded in 12 species from seven separate families**

Family	Species	Bird weight (g)	Nest weight (g)	Leaf petioles in S or L
Non-passerine families				
Cuculidae	*Coua ruficeps*	196.0	18.7	S
Passerine families				
Tyrannidae	*Myiopagis viridicata*	12.3	—	S
	Xenotriccus callizonus	11.2	—	L
	Colonia colonus	16.1	—	L
Furnariidae	*Thripadectes virgaticeps*	—	23.7	S
	Sclerurus albigularis	—	—	L
Formacariidae	*Formicarius nigricapillus*	62.0	26.0	L
Corvidae	*Nilaus afer*	18.7	9.7	S
Cinclidae	*Cinclus mexicanus*	61.0	300.4	S
Fringillidae	*Loxioides bailleui*	36.0	50.6	S
	Conothraupis speculigera	25.0	17.3	S
	Pinaroloxias inornata	12.5	18.5	S

S = structural materials; L = lining materials.
Weights of birds from Dunning (1993); weights of nests from nest survey.

to determine the origin of the silk, using the definitions outlined in Section 3.3. Silk was identified as one of five categories on this basis of thread diameter, waviness, the presence of double or multiple strands, and silk colour. This allowed it to be assigned to one of five categories: *spider web, spider cribellate web, spider egg cocoon, spider unknown structure*, and *caterpillar (lepidopteran)*. Table 5.4 shows that spider egg cocoon was the commonest single category of silk (35%) and that spiders were the originators of about 90% of the silk samples, compared with 10% from lepidoptera.

Silk-using species. Direct examination of nests and a review of the literature reveal the use of silk as a structural material in the nests of at least some species of 25 of the 45 passerine families (56%) (Sibley & Monroe 1990) and in one important non-passerine family, the hummingbirds (Trochilidae) (see Table 4.1). This supports the view expressed by Snow (1976): 'There are some materials on which many species of birds depend. Perhaps the most important is spiders web, which has unique qualities making it an ideal nest-building material...' with the reservation that *arthropod silk* is a more appropriate description than *spider's web*. The importance of silk is

Table 5.4. *Frequency of occurrence of arthropod silk in samples taken from the nests of six bird families*

	Trochilidae	Tyrannidae	Vireonidae	Corvidae	Sylviidae	Fringillidae	Silk type totals
Spider unclassified	6	12	0	13	1	6	38
Spider web	10	2	1	9	1	2	25
Spider cribellate web	1	5	0	3	1	0	10
Spider cocoon	5	13	5	13	2	8	46
Lepidoptera	1	2	2	2	2	4	13
Nest totals	23	34	8	40	7	20	132

particularly evident in the Trochilidae, in which 29 out of the 30 species examined built nests containing some silk. Snow (1976) further comments 'It is difficult to imagine what would have been the course of hummingbird evolution without spiders'.

Nest weight, bird weight and silk use. A one-way analysis of variance with bird weight and nest weight as dependent variables, and silk score (measured from 0 to 3 as described in Section 3.3) for the structural layer as a fixed factor, confirmed significant effects, with heavier birds tending not to use silk at all in their nests, and heavier nests not containing silk (Fig. 5.10, Hansell 1996a). The same analysis applied to the scores for silk as an attachment material showed the same effects for bird weight and nest weight (Humphries & Ruxton, unpublished analysis). It seems that, while silk is an excellent structural material for small nests (under about 30 g dry weight), it is of little value in larger structures, either for mechanical or economic reasons.

f) Grass
The scoring system described in Section 3.4 divides grass into four categories: grass leaf (narrow), grass leaf (broad), grass stem and grass heads. One or more of these categories was found in 37% of examined nests of Tyrannidae, 18% of Corvidae and 54% of Fringillidae. These different materials perform different tasks on the original plant and have different mechanical properties; therefore, to achieve a specific engineering solution, birds might require one of these four rather than another.

Grass leaf is a ribbon capable of a degree of bending on account of the tough but fine veins running in parallel along it. In the live grass plant, the leaves are designed to bend and, even dead, they can be curved to fit the shape of a nest cup. Grass stems, on the other

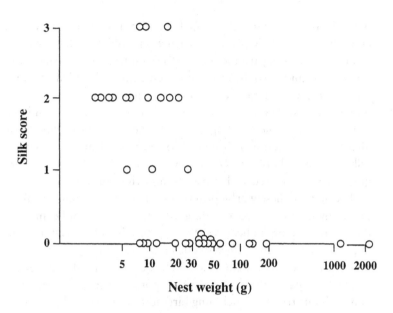

Figure 5.10
Plotting the amount of silk used in the structural layer of the nest (silk score) on a scale of 0 to 3 against nest weight for the Corvidae (Sibley & Monroe 1990), shows that it is smaller nests that are particularly dependent upon this material. (Adapted from Hansell 1996a.)

hand, are designed to support the leaves along their length and flowering heads at their apex. They are hollow beams, rather un-suited to being woven or easily curved. Grass heads are similar, but the fineness of the stems may allow substantial bending; as a cat-egory, they are quite variable since some grasses have very open flowering heads while others are more compact.

The use of flexible green grass leaf and strips of palm frond in the nests of the weavers (Ploceinae) (Collias & Collias 1984) shows that these materials are used for specialised construction. Building with grass heads is most clearly illustrated in the nests of the Passeridae, notably of certain Estrildinae, six of which were exam-ined in the survey. All the nests were wholly or largely composed of fine grass seed heads, tangled together to create a tight ball of swirling stems, as in the nest of the *red-cheeked cordonbleu (*Ura-eginthus bengalus*). In the nest of the *common waxbill (*Estrilda astrild*) the entrance itself is surrounded by a ring of outward-projecting seed heads, possibly forming some sort of mechanical protection. Grass heads are also used by certain weavers (Ploceinae) belonging to the same family. The nest of the *chestnut sparrow (*Passer eminibey*) is a ball of seed heads resembling that of *U. bengalus*, but in the *grey-headed sociable weaver (*Pseudonigrita arnaudi*) the nest is made by driving the stiff, narrow grass heads directly into the nest fabric, a process of thatching.

The specialised exploitation of dead grass stems as hollow beams is exhibited by certain warblers (Sylviidae), for example the *gar-den warbler (*Sylvia borin*), *blackcap (*S. atricapilla*) and *lesser whitethroat (*S. communis*), which build almost pure grass stem cups about 8 g in weight. The construction problem is that the grass stems resist bending around the small circumference of the nest cup, but the ingenious solution is that each stem is buckled at fairly regular intervals along its length to produce a polygon of short, hinged beams. These polygons are then stacked up around the builder to create the cup (Fig. 5.11). This construction principle is also found in the much heavier (50.5 g) nest of another Sylviid, *Sharpe's pied-babbler (*Turdoides (hypoleucus) sharpei*), where grass stems are mixed with grass leaf and vine tendrils.

The nests of these warblers may contain small amounts of silk or plant down which serve to stabilise the arrangement of beams, but the positioning of the beams across one another may itself have an important stabilising effect. Chilton, Choo and Popovic (1994) describe a type of three-dimensional beam structure, the *reciprocal frame*, used particularly as a roof support, and suggest that some animal-built structures including bird nests and beaver dams incorporate some of its features. The reciprocal frame has the constructional features of simple building rules applied to standard units in the form of beams. Structurally, the reciprocal frame has the merits that the completed array can span a distance greater than that of an individual beam, that the beams lock each other in position, and that stresses from localised loads are spread through the structure (Chilton *et al.* 1994, Fig. 5.12). However, a disadvantage is that, until the array of beams is complete, the whole structure is unstable; even after completion, removal of a single element may cause collapse. This accords with the apparent difficulty of twig nest builders such as the *rook in placing the initial twigs of the nest compared with placing those later in the construction. The problem of robustness of reciprocal frame structures is addressed by Chilton and Choo (1992), who suggest a number of solutions to reduce the risk of progressive collapse. Two of these appear to be simple options open to birds. These are, firstly, the projection of each beam beyond the intersection point with the beam on which it rests, resulting in intersection with other beams, and, secondly, duplication of the main beams and over-design of joints at intersections. Stick platform and grass stem nests should now be investigated in the light of these predictions.

g) Sticks
Sticks are beams and, in some species, nests are comprised simply of a platform of such beams. The family that best illustrates the stick platform design is the Columbidae (pigeons). The *pink-necked

Figure 5.11
The nest of the *blackcap (*Sylvia atricapilla*) is made of hollow grass stems, which are bent at intervals along their length to create polygons which, stacked upon one another, build up the nest cup. (Photograph by M. Kusmak.)

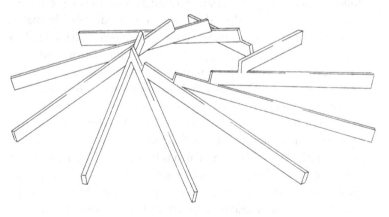

Figure 5.12
In a reciprocal frame roof beam array, each beam rests on another beam while having the preceding beam resting on it. (Adapted from Chilton *et al.* 1994.)

green-pigeon (*Treron vernans*) builds a platform of generally un-branched twigs of about 15 cm long and 3 mm diameter, on top of which lies a distinct second layer of finer (2 mm diameter) twigs of similar length. The *spotted dove (*Streptopelia chinensis*) (see Fig. 4.5) and *common woodpigeon (*Columba palumbus*) – nests of 43 g and 94 g in weight, respectively – make similar two-layered stick platforms of fairly standard-sized units. Structures of the same

form but larger size are made by species in other families, the
*striated heron (*Butorides striatus*) (nest weight 127 g), for
example, and the *sharp-shinned hawk (*Accipiter striatus*) (nest
weight 295 g) with fairly simple unbranched twigs of 15–25 cm
long and 3–5 mm diameter. The appearance of all these nests
viewed from below emphasises the tangential arrangement of the
edge twigs, suggesting a bird standing in the middle of the structure
laying out a succession of twigs in a standard orientation around
itself.

These platform nests are just the simplest in which stick beam
arrays are used. The nest of the *black-billed magpie (*Pica pica*) is a
deep twig cup reinforced inside with mud, often with an open twig
canopy protecting the brood from above. The nest of the brown
cacholote (*Pseudoseisura lophotes*) is a covered chamber approached by a horizontal entrance tube, all constructed of twigs
unsecured to one another, yet interlocked and robust (Nores &
Nores 1994a, Fig. 5.13). Evidence that these exhibit the properties
of a reciprocal frame, with the beams locking each other in position
and distributing localised loads throughout the structure, remains
to be demonstrated.

h) Design and convergence

The structural layer of a nest addresses the problem of strength and
cohesion in the chosen nest site. Examination of the major materials
involved has shown that there are certain identifiable material
solutions: silk and plant material to create a *velcro*, flexible fibrous
materials in tension or compression, and beam cups or platforms in
compression. These solutions may be expressed in a pure, un-
ambiguous form or in more hybrid structures; however, what is
repeatedly evident is that each can almost invariably be found in
more than one family. The same is true of architectural features.
One solution to siting nests on fine branches has been the top-lip-
supported hammock design, which is found in at least eight families
(see Table 5.2). The top or top-lip-attached hanging basket design,
created by 'weaving', appears to require some of the most complex
bird construction behaviour (Section 4.8). In spite of that, top-
attached nests held together by a 'weave and tangle' technique (that
is, without the aid of *velcro* or adhesive) were found in six passerine
families in the nest survey and the technique appears to occur in at
least four others (Table 5.5).

The occurrence of even relatively rare materials and design features in several families supports the view that, for many aspects
of nest building, there are only a limited number of successful

(a)

(b)

(c)

Figure 5.13
The construction of the
twig nest of a brown
cacholote (*Pseudoseisura
lophotes*) with a dome
and nest entrance tube.
(Redrawn from Nores &
Nores 1994a.)

solutions, and that nest building is sufficiently flexible under the influence of natural selection for most types of nest to have arisen independently more than once.

5.5 The nest lining

An important consideration in trying to understand the function of lining materials is selection pressure for small nests. Slagsvold (1982b) showed that fieldfares (*Turdus pilaris*) provided with larger nests had higher nesting success. This leads to the conclusion that, in certain circumstances, nest size may limit clutch size (Section 2.6). Slagsvold (1989a) identified three possible selection pressures that might favour smaller nests: the cost of nest building, predation risk, and enhanced rate of drying to reduce heat loss from the nest. The costs of nest building are considered in detail in Chapter 6, but no evidence points to selection for small nests to reduce building costs. Slagsvold (1989b) found no difference in the predation rates on *chaffinch (*Fringilla coelebs*) broods raised in larger rather than

Table 5.5. **Nests built as hanging baskets by 'weave and tangle'**

Family	Species	Bird weight (g)	Nest weight (g)	Nest outline
Eurylaimidae	*Smithornis capensis*	23.7	21.5	
	Corydon sumatranus	—	505.0	
	Cymbirhynchus macrorhynchos	—	88.0	
	Eurylaimus ochromalus	33.3	65.0	
	Calyptomena viridis	60.3	30.5	
Tyrannidae	*Tolomyias sulphurescens*	14.9	30.0	
	Pachyramphus niger	—	940.0	
Thamnophilidae	*Thamnophilus palliatus*	23.3	—	
	Carcomacra tyrannina	16.3	15.0	
Sturnidae	*Aplonis metallica*	61.0	450.0	
Passeridae	*Ploceus ocularis*	25.2	22.0	
	Ploceus vitellinus	33.6 ♂ 28.5 ♀	44.0	
	Malimbus scutatus	—	98.4	
	Anaplectes rubriceps	22.0	116.5	
Fringillidae	*Psarocolius wagleri*	214.0 ♂ 113.0 ♀	93.5	
	Gymnostinops montezuma	423.0 ♂ 225.0 ♀	36.4	
	Cacicus chrysopterus	39.0 ♂ 32.8 ♀	16.2	
	Icterus gularis	59.1	—	
	Icterus pustulatus	37.0	—	
	Icterus galbula	34.3	19.2	

The description 'weave and tangle' can be applied to 20 species from six passerine families in the nest survey. Published descriptions suggest that nests of species in other families fit this definition. Weights of birds from Dunning (1993); weights of nests from the nest survey. (Nest sketches by Jane Paterson.)
Species in other families where the published description appears to match that of hanging basket, weave and tangle.

Family	Species	Bird weight (g)	Reference
Furnariidae	*Cranioleuca erythrops*	16.9	Wetmore (1972)
Pardalotidae	*Sericornis citreogularis*	—	Campbell (1901)
Pomatostomidae	*Pomatostomus isedorei*	70.0	Coates (1990)
Certhiidae	*Troglodytes ochraceus*	9.4	Skutch (1960)

smaller nests; however, Møller (1990a), in an experiment in which blackbird (*Turdus merula*) nests of varying size were baited with artificial eggs, demonstrated that predation was positively correlated with nest size. This support for the predation argument may only apply to nests of smaller species, which may obtain additional concealment from minimising nest size. As argued in Section 5.4a, this may not apply to nests of larger species.

Predation may not be the only selection pressure for small nest size. Slagsvold (1989a, 1989b) proposed and demonstrated that smaller nests dried out more quickly, and concluded that this might lead to nests in more exposed sites being constructed with thinner walls of less absorbent materials. This, of course, depends upon the penalty to nest insulation of wet nest materials. However, Hilton *et al.* (in preparation) showed that the rate of heat loss from nest materials was dramatically raised by dampness of the materials.

The size of the nest is therefore affected by complex trade-offs in which larger nests may allow larger clutches and provide better insulation if dry, yet attract more predators and dry out more slowly if wet. One compromise solution to the predation problem is demonstrated by the ability of the nest cup of the *chaffinch to expand more than that of the *fieldfare during nestling growth (Slagsvold 1989b). This is because the *chaffinch nest is a *velcro* fabric (Storer & Hansell 1992), whereas that of the *fieldfare is mainly of grass, which resists expansion but is less water absorbent. Nest lining materials that duplicate some functions of the structural layer may be preferred because they do so more effectively and therefore reduce nest size; others may be chosen because they perform some function special to the lining of the nest.

a) The presence, number and type of lining materials

As defined in the nest survey (Section 3.3), lining materials are any layer of materials covering the inside of the nest cavity that have no clear structural role as judged from a non-destructive examination. The nest samples of Tyrannidae, Corvidae and Fringillidae showed that, in at least one-third of nests (respectively 35%, 38% and 33%), there were no lining materials (see Fig. 5.1). No relationship was found between the presence of lining materials and either cup diameter or nest weight (Humphries & Ruxton, unpublished analysis).

When lining materials were present, there was no clear pattern. Twenty-three different materials were recorded out of 32 categories. In the Tyrannidae, plant down and feathers were clearly the most frequently used materials (respectively nine and eight nests);

in the Corvidae it was grass stems and rootlets (respectively 12 and eight nests); in the Fringillidae it was fur/hair and rootlets (respectively 20 and 11 nests). In the whole nest survey, the ten most frequently occurring lining materials were, in order, grass stems, feathers, rootlets, plant down, fur/hair, palm fibres, broad leaves, fine woody stems, horsehair fungi, narrow grass leaves. This raises the question of their possible function or functions and whether there is such a thing as a specialist lining material. Of these materials, all are commonly found in the structural layer or as part of the outer nest layer, except feathers and fur/hair. Grass stems are of particular interest because of their popularity as lining material, while also being the second most frequently occurring structural layer material after silk. Silk was only once identified as a lining material, its strength possibly making it a danger to chicks, which might become entangled in it.

b) The function of linings

The selection of materials specially adapted for insulation can be seen as a method of minimising nest bulk while maximising heat retention. Although there is little known specifically about lining materials, the insulating properties of nests have been studied. Walsberg and King (1978a), studying two discontinuous incubators, the *red-winged blackbird (*Agelaius phoenicus*) and willow flycatcher (*Empidonax traillii*), found that the nesting energy expenditure of an incubating female in both species was 16–18% below that of a female perching near the nest. Even adding in the heat required to raise egg temperatures when the parent returns did not wipe out this reduction in nesting energy expenditure for the incubating bird compared with a perching one. Walsberg and King (1978b) found a similar difference for the *white-crowned sparrow (*Zonotrichia leucophrys*). The ability of the nest to buffer the clutch from fluctuations in external temperature may differ between species. Brown (1994) noted that the temperature inside the enclosed nest of the Cape weaver (*Ploceus capensis*) remained slightly above ambient when the female was in the nest. Zann and Rossetto (1991) found the insulative effect of the roofed nest of the zebra finch (*Taeniopygia guttata*) only to be evident when external night temperatures were particularly low, although the nest may also protect against extreme heat during the day.

In the rufous hummingbird (*Selasphorus rufus*) the cost of incubation has been shown to be related to cup depth (Calder 1973a). For an incubating bird to expose one-half rather than one-quarter of her body surface increases incubation costs by 41–62%. Smith, Roberts and Miller (1974) estimate that an increase of only 0.05 cm in the wall of hummingbird nests reduces energy expenditure on incubation by 13%. Skowron and Kern (1980) found that the thermal conductance of nests of selected North American songbirds measured between that of animal fur and that of wood, suggesting that they trap air, but not as well as fur. One of the better insulated nests of the 11 species studied was that of the *eastern white-crowned sparrow (*Z. leucophrys oriantha*), the nest of which has a fur lining; that of the *rose-breasted grosbeak (*Pheucticus ludovicianus*) has a rootlet lining and was among the poorest insulated of those studied. The insulative principle of these materials is trapped air, which may either be in the material itself (air spaces in mammal hairs) or in between the materials (feathers). Observing incubating *blue tits (*Parus caeruleus*) through a nest box video camera, it can be seen that they spend much time apparently fluffing up the materials around and below the eggs. The plant hair nest of Anna's hummingbird (*Calypte anna*) has to trade off the need to be compact enough for mechanical stability, yet trap enough air to have low thermal conductivity. In this it seems to be quite successful, its measured insulation being equivalent to that of polar bear (*Thalassarctos maritimus*) fur (Smith *et al.* 1974).

Møller (1984) tested certain predictions of the hypothesis that feathers have a special insulative role in the nest, and these were largely upheld. Early-breeding species of European passerines were more likely to have a feather lining to their nests than later breeders, and smaller species more likely than larger species. Nests of northern breeding species did not tend to have more feathers than those of southern species, although northern breeding citrine warblers (*Motacilla citreola*) were found to line their nests with feathers, whereas southern individuals did not. The results also upheld the hypothesis of a trade-off between the advantage of feathers in insulating the nest and the cost of increased nest detection by predators. Feathers were more likely to be used in enclosed nests, where they are less observable, than in open nests. The cost of feather lining in terms of increased predation risk was also confirmed experimentally by demonstrating higher predation on artificially baited nests with feather lining added compared to unfeathered nests (Møller 1987a). The benefits of feathers have, however, been shown by the effects of their removal compared with controls. In the nests of barn swallows (*Hirundo rustica*) this increased: rate of heat loss from eggs, incubation effort and nestling period (Møller 1991a); in nests of *tree swallows (*Tachycineta bicolor*), the same effects on incubation effort and fledging period were found, whereas fledging success was reduced (Lombardo *et al.* 1995).

Apart from incubation, the only other demonstrated function of nest lining materials is parasite control. Møller (1987b) observed that the number of feathers in the nest of the barn swallow (*Hirundo rustica*) rose during the egg laying and incubation phase and declined during the nestling period. He tested the possibility that feather removal might be a device either to reduce over-heating of the chicks or to reduce nest parasites. There was found to be no relationship between feather number and parasite load, leaving the changing requirements for thermoregulation as the most likely explanation of the change in feather number. Winkler (1993) found that the presence of feathers in the nest of the *tree swallow (*Tachycineta bicolor*) could more easily be related to temperature than to mite control, and Lombardo (1994) found in the same species that nests with larger numbers of feathers later in the season fledged fewer chicks, probably because of nestling hyperthermia in over-insulated nests.

Evidence that lining materials may be used to control arthropod ectoparasites is reported in Section 6.9, with good experimental evidence that secondary compounds in the fresh leaves of certain plants may be used for this purpose. The presence of fresh leaves was impossible to confirm in the nest survey, but horsehair fungus, which has reported antibacterial action (Melin, Wikén & Öblom 1947), did appear in the full nest survey as the ninth most common lining material (19 species of nest) and is known from published accounts as a lining material (McFarland & Rimmer 1996). However, it also appears in almost exactly the same number (21) of nests as a structural layer material. It may be a material with two separate roles, one structural, one antibiotic. However, more difficult to explain is the considerable popularity of grass stems. Their potential for use in hollow beam nests is explained in Section 5.4f (see Fig. 5.11), and they are second only in popularity as a structural layer material to silk, yet they occur more frequently as a lining material than any other. What is their function? It could be that the hollow stems provide good insulation, they may have unknown benefits in nest sanitation, or they may simply act as a comfortable surface to separate the eggs and chicks from the more abrasive structural layer.

CHAPTER 6

The cost of nest building

6.1 Introduction

The growing understanding of the energetic cost of reproduction in birds derives from studies on the costs of egg production, incubation and chick rearing, but there are, as yet, few such studies on nest building. This may in part be due to the difficulties in making the necessary measurements, but also to the undeclared assumption that nest building costs are likely to be low. This chapter pieces together such evidence as there is to test that assumption. For species that re-use nests or nest cavities, there appears frequently to be a cost to be borne of the accumulation of blood-sucking arthropods retarding chick development. This cost is borne as an alternative to building a new nest or, for cavity nesters, as an alternative to finding and competing for a new nest site. The second part of this chapter is an assessment of these costs and the behavioural responses of birds to minimise them.

The determination of construction costs and consideration of their implications have been best illustrated by research on silk-spinning arthropods. In such species, the energetic costs are those of the energetic value of the silk, the cost of silk synthesis, and of the construction behaviour. The cost of sheet-funnel web construction by the wolf spider (*Sosippus janus*) was determined from the oxygen consumption of spinning and non-spinning spiders in a respirometry chamber, plus the energetic value of the silk, determined by bomb colorimetry. This was found to be equivalent to more than 23 times daily standard metabolic rate (DSMR) (Prestwich 1977). The same technique applied to the construction of the sheet web of the linyphiid spider (*Lepthyphantes zimmermani*) showed that the total energy cost of web production was, at 10 °C, more than five times DSMR (Ford 1977). The respiration rate of caddis larvae (*Polycentropus flavomaculatus*, Trichoptera) when spinning a capture net was found to be 17% above resting level, and larvae fed on a limited diet were found to build nets with less silk (Dudgeon 1987).

The value of the investment influences other behaviour associated with building. Parsnip webworms (*Depressaria pastinacella*) attempt to find vacant silk shelters, or even try to usurp residents, rather than build their own. Shelter owners vigorously defend their sites against usurpers, whether they built them or won them. Measures of the cost of the shelter show that spinning it occupies one-third of a larva's time, requires 25% of the food not used in respiration and 18% of ingested nitrogen. The shelter is consequently a valuable resource (Berenbaum, Green & Zangerl 1993).

For larger organisms such as birds or mammals, measurements of energetic costs by respirometry may not be practicable. However, energetic costs can be estimated from time budgets observed in the wild, selective use of respirometry and the use of the doubly labelled water technique to determine energy costs for free living individuals. An example of such an approach, although not strictly of building, is the calculation of the costs of different kinds of locomotion in the golden mole (*Eremitalpa namibensis*). These locomotions are running on the dune surface and 'sand swimming' (burrowing) (Seymour, Withers & Weathers 1998). Respirometry was used to measure energetic costs of resting, running and burrowing, and field measurements of metabolic costs were obtained using doubly labelled water, while distances covered above and below the sand were obtained from the animals' tracks. Results revealed that sand swimming is 26 times the cost of running on the surface, but only one-tenth of the cost to mammals burrowing in compact soil. Similar comparisons in bird nesting await study, although preliminary attempts have been made to calculate energy costs of nest building directly. In addition, there are a variety of other kinds of indirect evidence to indicate something of the costs of nest construction and how natural selection has operated to minimise them. These include time taken to build nests, effects of differences in building effort on clutch size, various behavioural devices suggesting cost cutting, and the stealing of nests.

Species that nest in cavities not excavated by themselves have no construction costs, but the site itself has a value, which may be high since the birds themselves have no ability to create a site. In principle, the value of this could be estimated from the energetic costs of usurpation or of cavity defence, but such data are difficult to obtain. Tolerance of ectoparasites in such species therefore provides an indirect measure of the value of the site rather than of the cost of nest construction itself.

6.2 Calculation of energetic costs

A few studies have attempted to calculate the energetic cost of nest building directly. The cost of nest construction by *cliff swallows (*Hirundo pyrrhonota*) has been calculated as 122 kJ (Withers 1977). This is determined, for a bird of known weight and metabolic rate, on the basis that one collecting trip takes 20 s of flight at a known speed, 10 s of mud collection, and 30 s of working at the nest, and that to complete an average nest of 600 g takes about 1400 such trips. Collias (1986) carried out a similar calculation for nest building by male *village weavers (*Ploceus cucullatus*), based on a flight distance of a surprising 325 km. The energetic cost was determined as 158 kJ.

The energy expended by the cliff swallow was calculated to be similar to that of incubation, but less than the cost of raising the nestlings. Withers (1977) therefore concluded that the major cost of nest construction was in time rather than energy. Advancing the date of the start of egg laying is known to be important in some species. In the crested tit (*Parus cristatus*), if the male assists the female in the excavation of a tree hole cavity, the onset of egg laying may be brought forward by five days (Lens, Wauters & Dhondt 1994), which in turn improves the chances of fledglings entering winter flocks (Lens & Dhondt 1993). There is evidence here of significant energetic costs too. Help from the male in nest cavity excavation is dependent upon its body condition. A male ending the winter in poor condition apparently needs to give priority to foraging, in spite of the benefits to chick fledgings that its nest building efforts would provide.

Observations by Dolnik (1991) on the time needed for nest completion in 11 European bird species range between less than 24 hours for the yellow-billed magpie (*Pica nuttali*) and 200 hours for the penduline tit (*Remiz pendulinus*). For the penduline tit this is equivalent to 4.9 days basal metabolic energy costs, but for the osprey (*Pandion haliaetus*), for example, only 1.5 days. In the currency of egg production, costs of nest building in these 11 species can be expressed as equivalent to the energy spent on the synthesis of 0.5 clutches for the *willow warbler (*Phylloscopus trochilus*) and 2.7 clutches for the carrion crow (*Corvus corone*), respectively the cheapest and most expensive of the nests studied.

These estimations of energetic cost are based on metabolic rates obtained from respirometry studies on birds under laboratory conditions. Calculations of energy expenditure for free-living individuals over periods of several days can be obtained by using doubly labelled water. The energy costs are calculated from the decline of specific isotope activity in a blood sample taken at the end of the period, when compared with that in the initially injected water (Nagy 1980). For example, the cost of swimming in free-ranging jackass penguins (*Spheniscus demersus*) was calculated using doubly labelled water, knowing the duration and speed of swimming when foraging (Nagy, Siegfried & Wilson 1984). This technique could be applied to birds during the nest building period.

6.3 Gathering journeys and building time

The nest of the *woodpigeon *Columba palumbus* has about 200 twigs and weighs about 150 g (Møller 1982b). These twigs are typically gathered a small number at a time from the ground very close to the nest (personal observation). The total flight time, adjusted for the load carried, would give the energetic cost of building; that is, disregarding other possible costs such as increased predation risk whilst searching for materials and reduced escape performance when carrying them. This, for a nest only about 30% that of the weight of the bird itself, must be among the less expensive nests.

The weight of constructed nests relative to the size of the builder varies considerably (see Fig. 5.9), suggesting that construction costs may be similarly variable. The nest of the South American species, the brown cacholote (*Pseudoseisura lophotes*, Furnariidae) is composed of about 1000 twigs and weighs about 2.5 kg (Nores & Nores 1994a) – about 30 times heavier than the modest 86 g bird that builds it (see Fig. 5.13). More extreme still is the nest of the hamerkop (*Scopus umbretta*), a bird of less than half a kilo (420 g) (Dunning 1993). The nest is about 12.5 ft (3.8 m) in circumference in either the vertical or horizontal plane (Cowles 1930), composed of around 8000 sticks and weighing 'several hundred kilos' (Kahl 1967). At the other extreme, the nest of the crested treeswift (*Hemiprocne coronata*, Hemiprocnidae) is so small and fragile that the parent stands astride the nest to incubate the single egg (Lack 1956, Fig. 6.1).

Estimates of the time taken to complete a nest are similarly variable and, on the evidence available, bear no simple relationship to the absolute or relative weight of the nest. The *great grey shrike (*Lanius excubitor*) builds a robust nest close to the 65 g weight of the adult bird (Dunning 1993) in only nine days (Yosef 1992). The *black-billed magpie (*Pica pica hudsonia*) builds its domed twig nest, lined with mud and grass and about 20 times the weight of the bird, in an average of 43 days (Erpino 1968). Brown

Figure 6.1
The nest of the crested treeswift (*Hemiprocne coronata*) is a minimal structure. The bird stands astride the nest in order to incubate the egg. (Redrawn from Linsenmaier 1979.)

cacholotes, however, which build substantial twig nests as dormitories to be used out of the breeding season, can build 17 nests in a 36-month period, taking a mean of 23 days for each (Nores & Nores, 1994a).

Particularly long nest construction periods have been reported for swifts (Apodidae), where a large component of the nest material is the bird's own saliva. The 8.7 g nest of the black-nest swiftlet (*Collocalia maxima*), made of saliva and feathers, takes about 35–40 days, as does the wholly saliva nest of the edible-nest swiftlet (*Collocalia fuciphaga*) (Kang, Hails & Sigurdsson 1991). The little swift (*Apus affinis*) takes 1.8 months for adult pairs and a remarkable 4.6 months on average for yearlings (Hotta 1994). The Asian palmswift (*Cypsiurus balasiensis*), with its diminutive nest of saliva and plant down, takes a mere eight days (Hails & Turner 1984).

*Long-tailed tits (*Aegithalos caudatus*), birds of just 8.2 g (Dunning 1993), build nests of about 26 g (Hansell, in preparation), composed of about 3700 pieces (lichen, spider cocoon, moss and feathers). These materials may be collected more than one at a time and the work shared by the breeding pair. Considering the feathers alone, and assuming that they are collected about three at a time at a distance shown to average only 45 m (i.e. 90 m round trip) (Hansell, in preparation), then each partner would fly 22.5 km simply to line the nest. This helps to account for the construction of the nest taking about three weeks (Perrins 1979), whereas a *blackbird (*Turdus merula*) takes about three days (personal observation) to build a nest that weighs about twice the weight of the 113 g female (Dunning 1993).

The 28 g nest of the *bushtit (*Psaltriparus minimus*), a close relative of the *long-tailed tit, is a relatively large, domed nest about five times the weight of the builder, which is reported to take 13–51 days to complete, although the longer time seems to be due in part to interruptions caused by the weather. A more accurate measure of actual construction time is that observed for the hamerkop which, spread over six weeks and involving two builders in four hours of building a day, amounts to 336 hours of building (Kahl 1967). Nevertheless, a new nest is generally built each year (Cowles 1930).

6.4 Measurement of building costs as clutch reduction

Calculation of the energy costs of reproduction in terms of egg production equivalent does not simply emphasise that nest building may be a significant reproductive cost, but also that there may be a trade-off between energy expended on nest building and egg production. The nature of this trade-off may be revealed without experimental manipulation in species in which nest building costs vary greatly between individuals, although the pattern may be obscured by confounding variables such as bird quality. Most strikingly, this variation in individual effort occurs in species in which, in any breeding season, some birds construct new nests while others re-use one built in a previous year. *Eastern phoebe (*Sayornis phoebe*) nesting under bridges and in culverts in Indiana, USA, show such variation. The female builds a cup nest of mud which may be secured directly to the wall (*attached*) or perched on a ledge (*satant*), the former having a weight of more than 400 g, the latter weighing less than 200 g (Weeks 1978). Some birds, however, do not build a new nest at all, but *refurbish* one from a previous year. Breeding females therefore experience one of three levels of nest building costs, those building attached nests having the highest and those refurbishing existing nests having the lowest expenditure.

Weeks (1978) reported a near-significant and consistent trend for larger clutches in satant nest builders compared with attached builders, and larger clutches in nests of refurbishers compared to builders. However, Conrad and Robertson (1993) found some inconsistency in the application of the classes *attached* and *satant* and, on re-analysis of the data, found no difference between the two. They also found no difference in the time taken to construct attached compared to satant nests, contrary to previous claims (Klaas 1970). Building a new nest did take longer than nest refurbishment, but clutch sizes associated with each type of nest were indistinguishable. So, even where a difference in building effort was demonstrated, it appeared insufficient to affect clutch size.

In the *rook (Corvus frugilegus), nest construction also had no significant effect on clutch size when compared with refurbishment of a previous year's nest (Rutnagur 1990). Removal of all twigs from the local woodland floor before the start of the breeding season, in an attempt to raise the energetic cost of twig collection and so of nest construction, failed to reduce clutch size compared with control seasons, but did reduce the total number of breeding birds. A possible interpretation of this is that experimental removal of the twigs did raise building costs, but its effect on clutch size was to cause poorer quality birds to fail in their nesting altogether, so obscuring the difference between the 'twigs removed' and control years.

In the barn swallow (Hirundo rustica) the delay between the start of nest building and of clutch initiation was found to be greater for new nest builders compared to refurbishers (Barclay 1988). However, no difference in reproductive success measured as fledging success was found between the two groups. So, although energetic estimates suggest that nest building can be substantial, no subsequent effect on clutch size has yet been convincingly demonstrated.

6.5 Other evidence of nest building cost

A number of different kinds of evidence taken together do support the view that nest construction, either in terms of energy or time consumed, has been sufficiently important that selection has favoured the reduction of it. Examples are evidence of reduced nest size, the exploitation of the efforts of others through stealing nest materials or whole nests, and spreading construction costs over time.

The significance of minimal nests containing as little as one egg is uncertain; it might be either to reduce building costs or a consequence of predation pressure (Section 2.6) (Snow 1978). The other two types of evidence, although anecdotal, are more readily attributable to costs. In one study, between one-third and one-half of the nests of barn swallows and eastern phoebes located on concrete walls under bridges were found to be established on the nests of mud-daubing sphecid wasps Trypoxylon and Sceliphron

(Weeks 1977). The significance of this may not simply be a marginal saving on the cost of securing the nest to the wall, but the difference between success and failure of nest attachment. Indeed, Jackson and Burchfield (1975) attribute the expansion of the range of barn swallows into Mississippi to the nests of mud dauber wasps which are built under concrete road bridges. The generality of this technique is confirmed by a study in Ghana on the white-necked rockfowl (*Picathartes gymnocephalus*), whose roof-attached and wall-attached mud nests were found to be built on the nests of mud dauber wasps (Grimes & Darku 1968). This bird, at over 200 g Thompson (personal communication), is over ten times the weight of a barn swallow (18–20 g) and the eastern phoebe (19.8 g) (Dunning 1993) and probably builds the heaviest wall-attached mud nest of any bird (2 kg) (Glanville 1954). Perhaps this species would be unable to attach nests in its characteristic cave sites without the presence of mud dauber wasp nests.

In some bird species, the nest is begun long before it is due to be used, and grows episodically at times when building is cheap. This is illustrated by some species that build with mud and some that excavate burrows in soil, the hardness of which is affected by rain. The *rufous hornero (*Furnarius rufus*) builds its massive mud nests during the winter in bouts coinciding with mild weather and high rainfall when mud is most plentiful, and the blue-crowned motmot (*Momotus momota*) excavates its burrow when the soil is moist rather than when it is dry (Skutch 1964). The red-throated bee-eater (*Merops bulocki*) excavates its burrow at the end of the rainy season, although it does not nest in it till several months later (Fry 1972). Perhaps not surprisingly, the coincidence of burrow excavation and rain, presumably as an economy is also described for the sphecid wasp *Cerceris antipodes* in the arid regions of Australia (McCorquodale 1989).

Some species apparently spread the building costs over time by completing the building of the nest after the laying of the first egg, for example the chinstrap penguin (*Pygoscelis antarctica*) (Moreno, Bustamante & Viñuela 1995). This may not pose serious problems of supporting the eggs for these ground-nesting species but, more surprisingly the *bushtit (*Psaltriparus minimus*) may still add material to the nest after the laying of the first egg (Addicott 1938).

The cost of building a replacement nest after the failure of the first is minimised in some species by the salvage of materials from it to create a second. This is shown by the *chaffinch (*Fringilla coelebs*), which recycles spider silk (Marler 1956), and by the *long-tailed tit (*Aegithalos caudatus*), which recovers and re-uses the feather lining and probably silk as well (personal observation). Brown cacholotes (*Pseudoseisura lophotes*), each pair of which may build several stick nests during the year as dormitories, salvage twigs from one nest to build the next (Nores & Nores 1994a). This could constitute a significant saving as the average occupancy of a nest is only 42 days.

A similar behaviour to that of nest material recycling is that of stealing material from the nests of conspecifics. Five species of Tyrannidae are known to steal grassy nest material from one an-others' nests, particularly when nesting close together. They are three species of *Myiozetetes* (*similis, cyanensis, granadensis*), *Coryphotriccus parvus* and *Pitangus sulphuratus* (Smith 1980).

Nest material stealing is particularly associated with colonial nesting. Moreno et al. (1995) list 17 studies on more than a dozen colonial nesting species where nest material theft is reported, ranging from sea-birds such as frigatebirds (*Fregata*), gannets (*Sula bassana*) (Nelson 1978) to ploceine weavers (Crook 1964) and New World oropendolas (Icterini) (Skutch 1976). Chinstrap penguins, in common with other pebble nesting penguins, are habitual nest material thieves. The intensity of nest material collection and theft in this species is positively correlated with final nest weight, as is the intensity of nest defence by the male and female. The intensity of pebble theft is also greater in larger colonies, suggesting that it is exacerbated by shortage. This was confirmed by the experimental provision of stones to a colony, which resulted in a reduction in the proportion of stolen compared to collected stones (Carrascal, Moreno & Amat 1995). Stones that were stolen were larger than those collected, supporting the proposal (Burger 1974) that an additional advantage to stealing, beyond the reduction of construc-tion costs, is the acquisition of better quality materials. Larger chinstrap nests were found to be less susceptible to flooding, although this was not shown to influence fledging success (Moreno et al. 1995).

Stealing twigs from conspecific nests is a normal method of nest material gathering in the colonial nesting *rook (Corvus frugilegus)*. Stealing ability varies considerably between individual pairs, with the percentage of stolen sticks per nest ranging from only a few per cent to 30% (Ogilvie 1951) or over 50% (Rutnagur 1990). However, no correlation was found between the percentage of stolen nest material and either the rate of nest completion or clutch size, so the energetic benefits of stealing over collecting may be negligible in this species. The severest penalty may be in allowing your nest to be stolen from. Unattended nests disappear fast

(Ogilvie 1951, Rutnagur 1990). Stealing from neighbours, therefore, may be a tactic for a bird to be able to collect nest material
while still remaining able to defend its own nest.

A variety of species of birds from a number of families exhibit
facultative intraspecific brood parasitism. The occurrence of this
behaviour is particularly common in species which have
precocially developed chicks, such as ducks (Anatidae) and
pheasants (Phasianidae) (Yom-Tov 1980), although it is now recorded in a number of species that have altricial young (Rohwer &
Freeman 1989). The prevalence of conspecific nest parasitism in
species with precocial young supports the hypothesis that selection
against parasitism is light because the host pays only for the cost of
additional incubation, whereas in altricial species, selection for
nest vigilance will be greater since hosts will also pay an additional
cost of chick rearing (Rohwer & Freeman 1989). This hypothesis
takes no account of possible savings in nest-building costs by egg
dumpers, but individuals engaged in facultative conspecific nest
parasitism may at the same time be laying some eggs in a nest of
their own. An alternative hypothesis to females seeking to save on
incubation and chick-feeding costs is that egg-dumping females are
trying to salvage some reproductive success when presented with
failure through loss of their own nest (Eadie, Kehoe & Nudds
1988). Occasional intraspecific and interspecific nest parasitism in
some Iberian lark species (Alaudidae) is apparently due to high
nest losses through predation (Yanes, Herranz & Suarez 1996).
So, again, no advantage is attributed to savings in nest building;
however, the scarcity of nest sites could favour the parasitism of
the nests of others. This could explain the particularly high level of
conspecific nest parasitism occurring in cavity-nesting compared to
ground-nesting ducks (Eadie *et al.* 1988, Rohwer & Freeman
1989). Locating a nest to parasitise may also have costs, but these
should be low when neighbouring nests are easily accessible
(Yamauchi 1993).

6.6 Taking over the nest of another bird

Studies on conspecific nest parasitism have therefore contributed
little to an understanding of the value of the nest, but the adaptation
of second-hand nests and usurpation of nests from builders provide
examples where individuals apparently save building costs by their
behaviour.

Although natural selection might in general be expected to
favour diversification in the choice of nest site in coexisting species,

this situation might be reversed where nest sites are costly to construct, leading to competition for such sites as are available. In an analysis of data from a number of published nest usurpation studies, Lindell (1996) found that cavities and enclosed nests are more likely to be usurped than cup nests, and cavity usurpation is more likely to occur in temperate rather than tropical or subtropical zones. The higher incidence of enclosed nests in tropical habitats compared with temperate ones (Collias & Collias 1984) could, in the view of Lindell (1996), reflect the greater benefits in relation to cost for the more or less resident tropical nesting species compared to those for temperate nesters for which many species are migrants and there is a short breeding season.

Monk parakeets (*Myiopsitta monachus*) are unusual among parrots in making twig nests which may be joined together to create a collective structure (Forshaw 1989), but in Argentina they are reported to adapt nests of the firewood gatherer (*Anumbius citata*) (Humphrey & Peterson 1978). In another study in Argentina, half of 39 monk parakeet nests examined were former nests of the brown cacholote (Eberhard 1996). At least seven other parrot species are also known to adapt the nests of other birds. These include two species of lovebird (*Agapornis*) sometimes found in the nests of colonial weavers (Forshaw 1989), two species of lorikeet (*Vini* spp.), which may make use of the stick and grass nests of a variety of species, and three lovebird species (*Agapornis* spp.) which use nests of various African weavers (Forshaw 1989). The substantial nests of the osprey (*Pandion haliaetus*) may be adopted by a variety of species including, in North America, the great blue heron (*Ardea herodias*) and the *bald eagle (*Haliaecetus leucocephalus*) (Ewins *et al.* 1994).

These examples seem to be generally opportunistic and facultative, but the *solitary sandpiper (*Tringa solitaria*) is apparently dependent upon the nests of other species to breed. Field notes held at the Western Foundation for Vertebrate Zoology record twenty-three breeding attempts for this species nesting in Alberta, Canada, of which nine clutches were in nests of the *rusty blackbird (*Euphagus carolinus*, Fringillidae), six in nests of the American robin (*Turdus migratorius*, Muscicapidae), and the remainder in nests of other passerine species. The condition of the nests in this museum collection strongly suggests they were built in the same year as they were used by the sandpipers, although there is no indication that this species is a nest usurper.

The spot-winged falconet (*Spizapteryx circumcinctus*) is known to obtain monk parakeet nests for breeding, either by adoption or usurpation (Martella & Bucher 1984). Little swifts (*Apus affinis*) frequently take over the nests of other conspecifics and, when they do, they show conspecific infanticide. Hotta (1994) observed that on all eight occasions when a member of a breeding pair died, the chicks were killed when the replacement partner arrived. On virtually all occasions (11/12) when a new bird or a new pair usurped a nest, the acquired nest was more advanced than the one that had been left. In this species the metabolic cost of construction of the mucous nest is unknown, but the time cost is known to be substantial. Yearling pairs of little swifts take 4.6 months and even mature pairs take 1.8 months to complete a nest. Usurpation may therefore be a way of overcoming a substantial time and/or energy cost of reproduction. Evidence does, however, point to another possible advantage to an infanticidal bird. In all ten cases in which this was observed, the murderer obtained not only a nest but a partner that was older than the one they deserted. In an experiment in which *tree swallow (*Tachycineta bicolor*, Hirundinidae) males were removed from 15 breeding pairs, males took over in seven of them and the chicks were killed in five of these. This can be termed *resource competition infanticide* (Hrdy 1979), because the death of the young increases the resources available to the killer, in this case a female and a nest cavity.

Nest usurpation by non-excavating, cavity nesters probably reflects the value of the nest site since there is no cost of construction to be met. The scarcity of nest cavities is reflected in the reports of deaths resulting from attempted usurpations. House sparrows (*Passer domesticus*), when attempting to usurp *cliff swallows (*Hirundo pyrrhonota*) from their nests, may go to the lengths of destroying eggs or killing chicks (Samuel 1969). Occasional deaths of either *common starlings (*Sturnus vulgaris*) or woodpeckers of various North American species are reported to occur in struggles for woodpecker nest holes (Shelly 1935). Aggressive encounters occur between black-backed woodpeckers (*Picoides arcticus*) and *tree swallows (*Tachycineta bicolor*) which try to take over their nest holes (Short 1979). Such shortages may result in less competitive birds suffering lower breeding success or even in a proportion of adults being unable to breed at all. Skutch (1969) reports the *masked tityra (*Tityra semifasciata*) and black-crowned tityra (*T. inquisitor*, Tyrannidae) as repeatedly filling with leaves holes excavated by woodpeckers until the owners were forced to abandon them. Lack of available nest cavities for *prothonotary warblers (*Protonotaria citrea*) was found to contribute towards their acceptance of brood parasitism by the brown-headed cowbird (*Molothrus* *ater*). Desertion of parasitised broods by warblers was more likely in locations where nesting cavities were more abundant (Petit 1991).

6.7 The consequences of nest re-use

An important consequence of nest re-use and of colonial breeding using the same colony nest site is the build up of parasites that can re-infest the birds each breeding season (Møller 1994) Although birds play hosts to a great diversity of parasites including microbial infections, the most significant as far as the nest is concerned are populations of arthropod ectoparasites.

A variety of haematophagous arthropods are implicated: mites and ticks (Class Acarina) and also insects of the orders Hemiptera (bugs), Siphonoptera (fleas) and Diptera (flies). Their life cycles differ in certain respects, leading to important differences between them, particularly in the rate at which their populations can grow. The tropical fowl mite *Ornithonyssus bursa* found, for example, in the nest of barn swallows has three juvenile stages each lasting only four to five days. In each a blood meal is required before moulting. The eggs of the adult females laid in the nest re-infect the chicks, allowing a rapid build up of mite numbers. During the winter the mites can survive without feeding, awaiting the return of their blood suppliers (Møller 1990b, 1994).

Up to 45% of the first clutch nests of barn swallows may be infested with *Ornithonyssus bursa* mites by the time the chicks have fledged. The number of mites per nest is over 60, but the distribution is highly skewed for both first and second clutches, the majority having no mites at all but a few with as many as 10 000 or more (Møller 1994). Two thousand mites (nearly four/g of nest material) are reported from nest boxes of the naturally cavity-nesting tree sparrow (*Passer montanus*), although the boxes were parasite free at the start of the season. Mites of the genera *Acarus, Glycyphagus* and *Tyrophagus* were represented (Wasylik 1971).

The mud nests of the *cliff swallow harbour the swallow bug *Oeciacus vicarious*, a blood feeder as nymph and adult, and fleas (*Ceratophyllus* spp.) (Brown & Brown 1986). Fleas of this genus, notably the hen flea (*C. gallinae*), are common in the nest burrows of the sand martin (*Riparia riparia*) (Hoogland & Sherman 1976) and nest boxes containing great tits (*Parus major*) (Richner, Oppliger & Christe 1993).

Fleas are holometabolous insects having maggot-like larvae which feed on nest debris. Typically they over-winter in the nest cavity as pupae, emerging as blood-feeding adults in the spring. The generation time of the hen flea is only two to three days shorter than the time from chick hatching to fledging (18 days: Richner *et al.* 1993), so their capacity to increase their numbers within the host's breeding season is limited.

In a colony of the Peruvian booby (*Sula variagata*), where nests were a mean of 8 m apart on a guano island, nest desertions were observed to begin at one end of the colony and, over a period of a month, sweep across it. Desertion of the chicks was preceded by heightened levels of preening, found to be due to ticks of the genus *Ornithodorus*. The harvesting of guano in some areas of the island resulted in a recovery or breeding success for the boobies the following season due to the incidental removal of the dormant ticks (Duffy 1983).

Excluding adult biting flies (since they are not a specifically nest-related problem), the threat to nestlings from Diptera comes from flies of certain species which enter an occupied nest to lay an egg on the chick from which the larva, on emergence, burrows into the host. This, therefore, is not a problem of nest re-use, although it may be exacerbated by coloniality. In the New World tropics, botfly (*Philornis*) larvae cause the deaths of oropendola (*Psarocolius*) and cacique (*Cacicus*) chicks (Smith 1968b). The *pearly-eyed thrasher (*Margarops fuscatus*) does re-use nest sites, but there is no evidence that this enhances the infestation its chicks suffer from *Philornis* larvae (Arendt 1985a, 1985b). However, increase in the size of colonies of the barn swallow does appear to enhance infestation by blood-sucking blowfly larvae (*Protocalliphora*) (Shields & Crook 1987).

6.8 Indicators of the cost of nest re-use

The extent of the penalty of ectoparasite burdens resulting from nest re-use has been studied by comparing reproductive success in nests where parasites are absent with that in nests with known and, in some cases, manipulated numbers of parasites. For example, a comparison in the *cliff swallow between nests where swallow bugs *Oeciacus vicarius* had been removed by fumigation and untreated nests found no difference in clutch size, suggesting that the condition of the parents was not significantly affected up to that stage (Brown & Brown 1986). However, the greater the number of bugs per nest, the more the weight of the chicks was affected. In large swallow colonies, where swallow bug populations were found to be at their highest (Fig. 6.2), the mean weight of chicks at ten

Figure 6.2
Two 10-day-old cliff
swallow (*Hirundo
pyrrhonota*) chicks. The
bird on the left was from
a nest infested with
swallow bugs, the one on
the right from a nest
fumigated to remove
bugs. (Photograph by
Charles R. Brown.)

days in unfumigated nests was found to be 17% less than that of chicks in fumigated nests.

Møller (1990b), comparing the infestation of broods of *barn swallows from colonies in which all old nests had earlier been removed with that from unmanipulated control colonies, found no significant difference between them, either in terms of the percentage of nests with mites (20.9% experimentals, 23.3% controls) or in the number of mites per nest. This implies that new nests are primarily infected by parent birds, not by mites migrating in from neighbouring old nests. The number of mites per nest after fledging of the first brood was indeed strongly correlated with the mite load of adults arriving in the spring (Fig. 6.3).

The effects of mite level on brood were estimated by comparing new nests from which mites were removed by pyrethrin spray with unsprayed new nests and with sprayed nests to which 50 mites had then been experimentally added. Fewer chicks were reared in the mites-added nests and the body mass of the chicks was also about 5% lower in both first and second clutches. Nestlings reared in the mites-added nests had a shorter nestling period than those in the sprayed-only nests, and it was found that the shortened nestling period adversely affected subsequent survival. The early fledging of the mites-added clutches could possibly have allowed parents of this group to start their second clutch earlier than those of the sprayed-only group. In fact, the interval was 8% longer, indicating that the body condition of the mites-added parents was also adversely affected. Females in the mites-added nests or

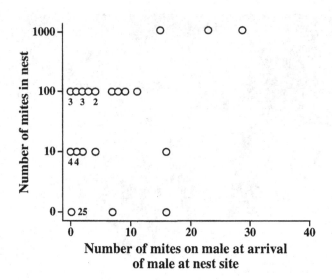

Figure 6.3
A positive correlation
shown between the
number of mites in barn
swallow (*Hirundo
rustica*) nests following
fledging of the first brood,
in relation to the number
of mites recorded on the
head of a male barn
swallow nest owner upon
arrival in spring.
Numbers beside or below
data points refer to the
number of birds included
at that point. (Adapted
from Møller 1990b.)

unsprayed nests laid significantly smaller second clutches than those in the sprayed nests, and were unlikely to attempt a second clutch at all if the infestation was high (Fig. 6.4, Møller 1990b).

A study comparing the breeding success of the cavity-nesting *purple martin (*Progne subis*) in colonies infested with the mite *Dermonyssus prognephilus* with that of birds whose nests were treated with acaricidal dust showed similarly marked effects. In untreated colonies, parents were more likely to desert the nest, chick weight and fledging success were reduced, as was the likelihood of survival after fledging (Moss & Camin 1970).

The effects of flea infestations have also been shown to be significant. The breeding performance of great tits (*Parus major*) in nest boxes harbouring *Ceratophyllus gallinae* was compared with that of pairs breeding in nest boxes sterilised before the start of the season in a microwave oven (Richner *et al.* 1993). Untreated boxes after chick fledging contained an average of 37 fleas, but infestation had no effect on clutch size or hatching success when compared with the sterilised box group. The chicks in the treated boxes gained most body mass regardless of clutch size, whereas the body mass of chicks from infested boxes either failed to rise or decreased (Fig. 6.5). The nutritional condition of chicks (measured as body mass/tarsus length) was also found to be sharply reduced in the parasitised group at 17 days as the chicks were nearing fledging.

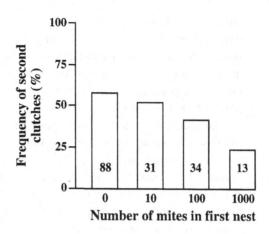

Figure 6.4
Frequency of second clutches of barn swallows (*Hirundo rustica*) in relation to the estimated number of mites in the nest used for the first clutch. Numbers within the bars refer to numbers of barn swallow pairs. The frequency of second clutches tends to show a decrease with an increase in the number of mites in the first nest. (Adapted from Møller 1990b.)

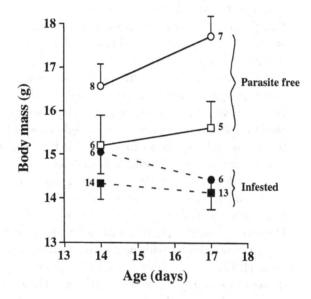

Figure 6.5
The effect of flea infestation on the growth of great tit (*Parus major*) chicks. Mean body mass (+/– 1 S.E.) of nestlings 14 and 17 days after hatching. Circular symbols represent small brood sizes; square symbols represent large brood sizes; open symbols, parasite free; closed symbols, infested. (Adapted from Richner *et al.* 1993.)

6.9 The response of nest re-users to blood-feeding ectoparasites

The build-up of haematophagous ectoparasites through nest re-use has led to the evolution of behaviour to avoid or minimise its consequences. The setts of the European badger (*Meles meles*), which may survive for decades or longer, can grow to contain hundreds of metres of burrow and dozens of nest chambers, yet only be occupied by a handful of badgers (Roper 1992). Badgers within a sett tend to change their bed chamber every two or three days and avoid one slept in by another badger on the previous day. Badgers treated to remove fleas, ticks and lice show less tendency to move sleeping quarters, supporting the hypothesis that the behaviour is adapted to control ectoparasite populations in the sett (Butler & Roper 1996).

The response of some bird species to infested nests or nest cavities is to avoid re-using them. Barn swallows in a Canadian population were found to avoid mite-infested old nests in favour of building a new nest for the first clutch. A large proportion of birds at the site favoured the cost of nest construction over the penalty of nest re-use. Any costs of nest rebuilding seem to have been compensated by the avoidance of ectoparasites because reproductive success proved to be the same for re-users and rebuilders (Barclay 1988). However, the breeding season in this location is unusually long due to extended insect abundance. Barclay (1988) suggests that the cost of nest reconstruction is largely one of time delay in starting breeding and that at this site the cost is low due to a longer breeding season.

Great tits will choose a flea-free rather than a flea-infested nest box (Oppliger, Richner & Christe 1994). This should not, however, be seen as specifically a reproductive adaptation, as they also show the same discrimination in choosing night roosts outside the breeding season and may choose to roost out in the open rather than in an infested box (Christe, Oppliger & Richner 1994).

A study in Spain showed that European pied flycatchers (*Ficedula hypoleuca*) discriminated against nest boxes containing last year's material in favour of clean ones (Merino & Potti 1995). However, a similar study conducted in Sweden found that they preferred nest boxes with nest material and did not discriminate between material with or without ectoparasites. The conclusion drawn in this study was that, with the short breeding season, birds need to make quick decisions on the acceptability of a nest cavity. In these circumstances, evidence of occupation in a previous season may be an indicator of nest site quality (Olsson & Allander 1995).

In *cliff swallows nest reconstruction is more common in large than in small colonies because the level of swallow bug infestation is positively correlated with colony size (Brown & Brown 1986). However, individual pairs may reduce the cost to their clutch of re-using a parasited nest by placing eggs in the nest of close neighbours which have low parasite burdens (Brown & Brown 1991). If such a dumped egg is to be accepted, it must be introduced at about the same time that those of the host are being laid. To do this, the egg-dumping female will either lay an egg directly into that of a neighbour or, more remarkably, transfer one in her beak within a few days of her laying it in the home nest. Egg-dumping females can evidently judge the state of infestation of nests near their own. Selected nests were found to be more likely to fledge young that those that were ignored.

Barn swallows are apparently unable to predict the effect of mites in the nest on the fledging success of their chicks. Although an infestation of 10 000 mites reduces the most productive clutch size from 8.3 in the first clutch down to 5.9, parents make no adaptive adjustment in clutch size (Møller 1991b). However, Møller (1994) points out that, provided the resources available for reproduction are not fixed, parents could compensate for those diverted by the parasites through extra foraging effort. Evidence of this comes from reproduction in the house wren (*Troglodytes aedon*); nestlings in broods hosting more than ten blowfly larvae (*Protocalliphora parorum*) and hundreds or thousands of mites (*Dermanyssus hirundinis*) per nestling did not experience higher nest mortality, nor did they have their growth much depressed compared with chicks in parasite-free nests, even though the parasites consumed an estimated 10–30 g of blood per brood by the time the nestlings had fledged (weight of adult bird = 10.9 g; Dunning 1993, Johnson & Albrecht 1993). The parents, it seems, compensate for the diversion of resources by increasing their own energy expenditure in food gathering.

In addition to responding to nest parasitism in the above ways, it appears that some species of nest re-users attempt to control the ectoparasite population. The most widespread method of doing this appears to be through nest fumigation using the secondary compounds present in some plants which are toxic to arthropods, although other hypotheses for the collection of green plant material have been proposed.

Johnston and Hardy (1962) noted that the cavity nesting purple martin, after completing a nest of dead leaves and sticks, placed green leaves round the edge of the cup, thereby possibly enhancing nest humidity, or killing either arthropods or bacteria through the release of hydrocyanic acid (HCN) on rotting. More recently, it has been proposed that it is the volatile secondary compounds present in some plants that are the active agents. Additional hypotheses for the behaviour have been that the green materials provide shade, or possibly insulation (Clark 1990), concealment or, indeed, social advertisement (Bucher 1988).

Confirmation of the fumigation hypothesis requires the following evidence. Firstly, that species showing the behaviour are those most likely to be affected by nest parasites. Secondly, that birds collect green plant materials preferentially from species that contain compounds toxic to arthropods. Thirdly, that these materials do, indeed, reduce parasite numbers in the nest. Evidence of all these has now been gathered. A significant positive association between nest re-users and nest fumigation was demonstrated for 49 species of North American and European birds of prey (Wimberger 1984). Of 28 species known to incorporate fresh greenery into the nest, 22 re-use their nests, whereas only eight of 20 species that use no greenery also re-use their nests, a significant positive relationship. A similar association between nesting habit and nesting materials was found in a study of 137 North American passerines (Clark & Mason 1985). Half of the secondary cavity nesters incorporated green plant materials into their nests, whereas cup nesters, which rarely re-use nests, were unlikely to make use of the fresh plant material (Fig. 6.6). Nevertheless, the association between nest re-use and the collection of green plant materials is apparently far from universal. Monk parakeets (*Myiopsitta monachus*), which use their communal nests as dormitories, may use fresh green material in the nest at the time of breeding, although the evidence for this is unclear (Bucher 1988). The buff-throated foliage-gleaner (*Automolus ochrolaemus*, Furnariidae), which nests in burrows, does line its nest cavity with green leaves of a mimosa species (Wetmore 1972); only four out of 68 other cavity or enclosed nesting Furnariidae and Dendrocolaptidae are reported to do the same. It remains to be seen whether careful study of these species will increase this low number.

Studies on single bird species provide additional evidence of the function of green plant materials in the nest. The first is evidence that the birds select only certain species of plant, the second that the plants chosen do control arthropod ecotparasites and even bacteria. House sparrows (*Passer domesticus*) nesting in cavities in buildings in Calcutta incorporate into their nests leaves of the neem tree (*Azidirachta indica*), a practice shared by local people who traditionally place fresh neem leaves in their cupboards to protect clothes from insect damage (Sengupta 1981).

The *common starling (*Sturnus vulgaris*), a typical cavity nester, takes a variety of green plant materials to the nest hole prior to egg laying, but does not simply gather them in proportion to their availability. One favoured herb, *Agrimonia paraflora* (Rosaceae), is capable of inhibiting bacterial grown on a nutrient medium (Clark

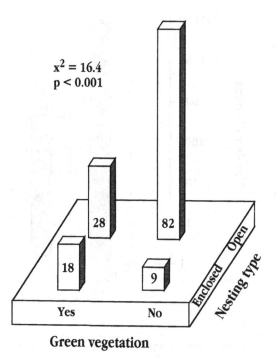

$x^2 = 16.4$
$p < 0.001$

28   82

18   9

Open

Enclosed   Nesting type

Yes   No

Green vegetation

Figure 6.6
The frequency
distribution of nesting
type (open cup nests vs.
enclosed spaces) and
whether or not the species
incorporates fresh green
vegetation into the nest.
The numbers represent
the numbers of North
American species
surveyed which fell into
each of the four
categories. (Adapted from
Clark & Mason 1985.)

& Mason 1985). The volatiles in some other plants collected by starlings are more effective in retarding the hatching of louse eggs (*Menacanthus* sp.) than volatiles from a selection of plants which they ignore. One selected herb, *Solidago rugosa* (Compositae), contains the sequeterpenes, a-bornyl acetate and farnesol, both known to act as juvenile hormone analogues, retarding insect development (Rosenthal & Janzen 1979). The ability of these birds to discriminate between effective and ineffective plants is olfactory (Clark & Mason 1987), but is apparently seasonal, developing at the start of the breeding season and being lost by the end of the summer (Clark & Smeraski 1990).

Other plant species selected by the common starling have been shown experimentally to be active against arthropods. For example, wild carrot (*Daucus carota*, Umbelliferae) added to the nest reduces the number of mites in starling nest boxes compared to those of untreated controls (Fig. 6.7, Clark & Mason 1988, Clark 1990). Chicks raised in nests with added wild carrot leaves also maintained higher blood haemoglobin levels, although feather production did not differ between them and controls.

Treating samples of infested starling nest material in the laboratory with leaves of wild carrot or of *Erigeron philadelphicus* (Compositae), another plant selected by starlings and significantly known

Figure 6.7
The effect of wild carrot on the number of mites in starling nests at the time of chick fledging. Treatments were: PR = no green plants in the nest; PA = addition of wild carrot leaves every seven days; NR = replacement of all nest material every seven days. Data show significant treatment and time effects on the mite populations. (Adapted from Clark 1990.)

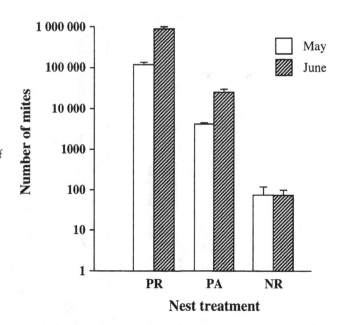

as fleabane, suppressed mite numbers compared with untreated control material. Leaves of garlic mustard (*Alliaria officinalis*, Cruciferae), a species not selected by starlings, left mite numbers in nest material the same as those in untreated control material. The effects of wild carrot and fleabane seem not to be due to their ability to kill mites but to their ability to suppress moulting and therefore delay mites reaching maturity (Clark 1990). Because natural population growth in these mites can be very rapid, retardation due to these plant volatiles can have very substantial effects. For example, starling nests from which green plant material was removed during the second breeding attempt had an average of 780 000 mites after 21 days, compared with 8000 mites in the unaltered group. From this it is estimated that a delay of as little as three days in the onset of normal population growth could reduce daily blood loss per chick from 10% per day (even as high as 40% per day) down to just 1% (Clark & Mason 1988, Clark 1990).

Although the hypothesis that green leaves in the nest are used to control parasitic arthropods is supported by a full range of evidence, rival hypotheses cannot be entirely dismissed. Some evidence for green plant materials as nest camouflage, for humidity control or for advertisement of nest occupancy is provided by one or two species (Bucher 1988). In the case of the starling, it is hypothesised that these materials could be used as a courtship display; this is supported by their collection being largely a male behaviour and more pronounced in the lining of new rather than re-used nest boxes (Gwinner 1997). The collection of green plant nest materials can be imagined as evolving from the benefits of bringing green plants to build the nest rather than the normal dead material. This innovation could then have spread through the population as an inherited or possibly even as a learned trait.

More surprisingly, a similar selective advantage may have resulted from the modification of behaviour in which prey captured for chicks of the eastern screech owl (*Otus asio*) are brought to the nest. Parent owls bring all prey to the nest dead except for blind snakes (*Leptotyphtops dulcis*). Some of these blind snakes may be killed and eaten by the chicks, but others escape into the debris at the bottom of the nest cavity. In a study site where 18% of screech owl nests were found to be occupied by blind snakes, chick growth rate and survival were found to be greater in nests containing live blind snakes than in those without (Gehlbach & Baldridge 1987). It is not clear how the owl chicks benefit, but among the arthropods colonising the nest debris are dipteran larvae; control of these by blind snakes may protect the owl chicks from their attacks.

CHAPTER 7

The selection of a nest site

7.1 Introduction

Koepcke (1972), in a study of the nest sites of birds in Peruvian rainforest, describes nests in the canopy of large forest trees (lined forest-falcon, *Micrastur gilvicollis*), low down in thickets (hoatzin, *Opisthocomus hoazin*, and *great antshrike, *Taraba major*), the ends of fine branches (sulphur-rumped flycatcher, *Myiobius barbatus*), and the ends of leaves (little hermit, *Phaethornis longuemareus*). Species belonging to four different families (Hirundinidae, Picidae, Psittacidae, Tyrannidae) were nesting in tree holes, each of characteristic location, and species of a further six families were nesting in burrows in the ground (Alcedinidae, Buccionidae, Furnariidae, Galbulidae, Hirundinidae, Momotidae), one species of quail (Odontophoridae) was nesting in cavities in the leaf litter, and a puffbird (Bucconidae) was nesting in an arboreal termite nest (Fig. 7.1). These species-typical nest sites raise questions concerning, firstly, the factors which influence nest site selection and, secondly, the adaptive significance of the properties of the chosen site. These are the questions examined in this chapter.

The use of an upwardly directed fish-eye lens showed that the nest of the warbling vireo (*Vireo gilvus*) in Arizona was sited so that two to four times more sky was visible through the canopy foliage in the eastern than in the western portion of the upper hemisphere. This had the effect of exposing the nest to 90% more direct sunlight in the morning than in the afternoon, apparently to reduce the thermal stress on the brood in the afternoon (Walsberg 1981). Nest site selection clearly goes beyond locating branches with a suitable geometry for nest attachment.

Nest site selection in the whimbrel (*Numenius phaeopus*) in Canada was found to favour hummock-bog rather than two similar neighbouring habitats in which it experienced lower breeding success. Nests were also characteristically located within 1 m of a shrub. This was no closer to a shrub than sites chosen at random within the hummock-bog habitat, but in the other two habitats, where shrubs were more scarce, selection of sites close to a shrub was clearly apparent (Skeel 1983). This behaviour seems to be a nest concealment adaptation to predation, which is the primary cause of egg loss. However, nests were also preferentially located on hummocks, probably to raise them above the waterlogged ground. So, nest site selection in this species has adapted to both biological and physical selection pressures on its breeding success.

Through nest construction, birds create a highly localised environment to their own specifications. This shapes the climatic conditions for eggs and, in many species, chicks as well. But nest construction is preceded by nest site selection to ensure that the nest itself is located in a wider environment, possessing attributes contributing towards fledging success. Four factors which influence that choice are apparent. Two of these (climate and predators) have already been mentioned. The third is availability of food to raise the chicks, which is not considered in detail here because it has no direct bearing on the success of the nest itself. A fourth possible influence on the selection of a site for the nest is the availability of suitable materials to build it.

7.2 The availability of nest material

Birds typically make use of local materials. This may have led to the conclusion that locating materials poses no particular problems. However, for species that are nest material specialists, the choice of a nest site could be influenced by the availability of the right material at a critical time. Little consideration has been given to this problem and there is no experimental evidence to indicate that it is even true. There are, nonetheless, indications that it deserves investigation.

The concept of nest material availability as a constraint on nest site selection and even on breeding range was first clearly articulated by Tomialojc (1992). He was struck by the absence of breeding song thrushes (*Turdus philomelos*) but presence of *blackbirds (*Turdus merula*) in drier Mediterranean habitats and some more northerly urban areas. In terms of food supply for chicks, thrushes should be as well off in these areas as blackbirds, and seem to be breeding well in neighbouring regions in the presence of blackbirds. Tomialojc (1992) proposed that in dry and in urban habitats the thrushes failed to breed because of the lack of the combination of wet conditions and rotten wood, or substitute materials like mud or dung, needed to create the typical nest lining. The composition of nest linings in different parts of their range tended to support this hypothesis.

Figure 7.1
Examples of the variety of nest sites utilised in Peruvian rainforest. Large
forest trees (a) the lined forest falcon (*Micrastur gilvicollis*, Falconidae).
Bushes and thickets (b) the hoatzin (*Opisthocomus hoazin*, Opisthocomidae),
and (c) the *great antshrike (*Taraba major*, Thamnophilidae). Holes in trees
(d) martins (*Progne*, Hirundinidae), (e) woodpeckers (*Melanerpes* and
Dryocopus, Picidae), (f) *tityras (*Tityra*, Tyrannidae), and (g) parakeets
(*Aratinga*, Psittacidae). At the end of a hanging vine (h) the sulphur-rumped
flycatcher (*Myiobius barbatus*, Tyrannidae), or end of a leaf (i) the little
hermit (*Phaethornis longuemareus*, Trochilidae). In burrows in banks,

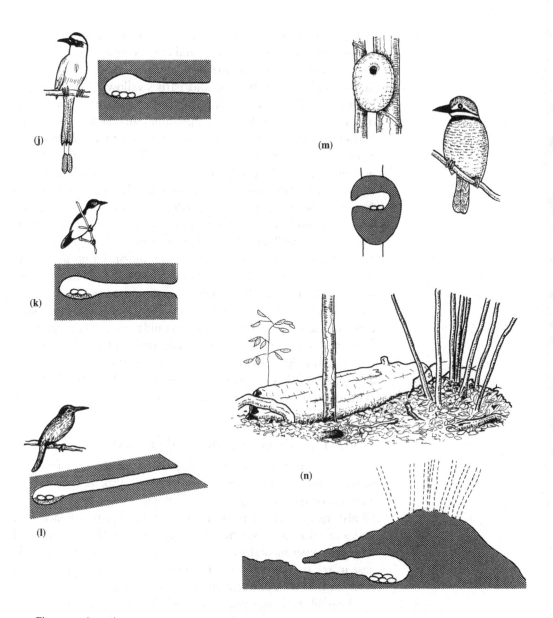

Figure 7.1 (*cont.*)
(j) the blue-crowned motmot (*Momotus momota*, Motmotidae), (k) the
olive-backed foliage-gleaner (*Automolus infuscatus*, Furnariidae) and (l) the
bluish-fronted jacamar (*Galbula cyanescens*, Galbulidae). Cavities in arboreal
termite nests (m) the chestnut-capped puffbird (*Bucco macrodactylus*,
Bucconidae), or chambers in the leaf litter (n) the starred wood-quail
(*Odontophorus stellatus*, Odontophoridae). (Adapted from Koepcke 1972.)

Chapters 4 and 5 have already shown that selection of nest material is often species specific and may be more specialised than we appreciate. The example of differences in the proportion of sand in the mud selected by cliff and barn swallows (*Hirundo pyrrhonota* and *H. rustica*) illustrates how an apparently unitary class of material may embrace recognisable sub-classes. Storer and Hansell (1992) showed that nests of the *chaffinch (Fringilla coelebs) contained mainly spider cocoon silk, and that the web silk was all of the cribellate rather than ecribellate type. The *long-tailed tit (Aegithalos caudatus) is a specialist in the choice of at least four of its materials, choosing lichens predominantly of only two genera (*Parmelia* and *Physcia*), spider cocoons of *Lyniphia triangularis* or similar type, small-leafed branching mosses, typically *Eurynchium praelongum* and *Hypnum cupressiforme*, and bird contour feathers; it also requires them all in large numbers (about 3000 lichen flakes, probably more than 600 cocoons, 200–300 sprigs of moss and about 1500 feathers: Hansell 1993b, in preparation). The coincidence of adequate quantities of these could therefore be a significant factor in nest site selection or even in the timing of breeding. Hansell (1995) found that the occurrence of feathers was patchy and scarce, and suggested that long-tailed tits might be early nesters in part to avoid competition from other species for feathers or spider cocoons.

7.3 The influence of physical factors

The control of the incubation temperature of the eggs is a clear influence on nest design (Section 5.5) and is also, as shown in the example of the warbling vireo, an influence on nest site selection (Walsberg 1981). Evidence of the importance of physical factors on nest site selection nevertheless remains essentially observational, with few experimental studies on the physical environment of the nest in relation to nest site, or comparisons of breeding success between sites with different climatic conditions.

The published examples are overwhelmingly concerned with the effect of nest site on brood temperature. The *white-crowned sparrow (Zonotrichia leucophrys) is, like the warbling vireo, a species that apparently regulates nest temperature by selecting a suitable pattern of overhead vegetation (Walsberg & King 1978b). *Broadtailed hummingbirds (Selasphorus platycercus) nesting at about 3000 m in the Rocky Mountains typically place their nests either low down in spruce (*Picea*) or fir (*Abies*) and sheltered by an overhead branch or, in aspen (*Populus*), near the canopy but under branches (Fig. 7.2). Using eggs containing thermistors in artificial

nests, Calder (1973b) found that nests in preferred sites lost less heat through convection, conduction or radiation than those at sites without overhead protection. Similar protected sites are chosen by the smallest North American bird breeding north of Mexico, the *calliope hummingbird (*Stellula calliope*) (Calder 1971). This site choice could be a critical adaptation against the lowest local temperatures, which are reached in the hours before dawn.

Other species choose the side of a bush which faces in a particular direction. The superb fairywren (*Malurus cyaneus*) chooses the northern sector of a bush, the white-winged fairywren (*M. leucopterus*) the north-east (Tidemann & Marples 1988). The *pinon jay (*Gymnorhinus cyanocephalus*), which nests in the canopy of pine trees, overwhelmingly (83%) selects the southern sectors of the tree, apparently to allow additional sunlight to reach the nest, saving incubation costs in this early nesting species (Balda & Bateman 1973).

Sage sparrows (*Amphispiza belli*) in Idaho place their nests on average 34 cm above ground, 21 cm from the shrub perimeter, and avoiding the south-west sector from which not only comes the hot afternoon sun but also periodic strong winds (Petersen & Best 1985a). The nests of the western hermit thrush (*Catharus guttatus*) in Arizona are almost always (93%) located in firs (*Abies concolor*) of 1–3 m tall. These typically grow in lower drainage areas which produce additional soil moisture. The nests are also preferentially placed in the south-west sector of the tree (Martin & Roper 1988).

Changes in preferred nest site over time, possibly in response to the changing physical environment, have been observed in the

Figure 7.2 Broad-tailed hummingbirds (*Selasphorus platycercus*) nesting at about 3000 m place the nest so that there is shelter from an overhead branch. (Adapted from Calder 1973b.)

*yellowhammer (*Emberiza citrinella*), for which ground nests are commoner in April than later in the season (Peakall 1960), and in Brewer's sparrow (*Spizella breweri*), which also shows a slight increase in the height of the nest above ground as the breeding season progresses (Petersen & Best 1985b). Clearer evidence that such trends regulate nest temperature is shown for the rufous hummingbird (*Selasphorus rufus*) in British Columbia. Horváth (1964) showed that the lower spring nest sites were more protected against climatic extremes than high ones. In the summer, however, high radiation levels make temperature fluctuations nearer to the ground greater. At this time of year the hummingbirds place the nests higher up, but in the shelter of deciduous leaves which provide a similar nest microclimate to that selected in the spring.

7.4 The influence of predators

Evidence of the influence of predation on the choice of nest site, in contrast to that of either physical factors or nest material availability, is extensive and detailed, so it provides the strongest argument that survival of the brood to fledging may depend on the location of the nest itself. MacArthur's (1958) demonstration that five co-existing species of warbler differed in their foraging locations in spruce trees appeared to support the view that feeding competition was the reason for their separation, until Martin (1993c) showed, on more detailed examination of the same data, that their foraging locations in fact overlapped considerably. The foraging and food competition hypothesis also seemed to be supported by data showing that increasing numbers of species co-exist in habitats with greater structural variety and therefore apparently more types of foraging sites (MacArthur, MacArthur & Preer 1962, Willson 1974). However, Martin (1988b), reworking Willson's data with birds classified on the basis of both foraging and nesting sites and adding field data of his own, showed that it was easier to explain increasing numbers of co-existing species in terms of the vegetation layer in which they nested rather than that in which they foraged (Fig. 7.3). It does, therefore, seem that the selection of the nest site can be explained in terms other than the distribution of specialised feeding sites. The rest of this chapter confirms the importance of nest predation in that selection process.

Ninety-three per cent of the nests of the western hermit thrush are located in small white fir trees (*Abies concolor*), although the bird does not utilise these trees for foraging (Martin & Roper 1988). The location of nests in the south-west sector of firs has

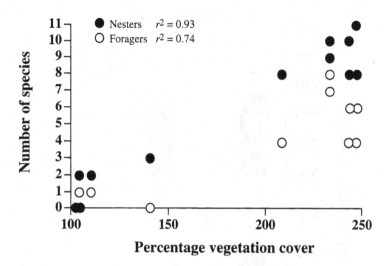

Percentage vegetation cover

Figure 7.3
The number of species of off-ground nesting birds in relation to percentage vegetation cover in woodland habitats in Illinois. When plotted as the number of species *nesting* (solid circles) a stronger correlation is provided than when plotted as species *forgaging* (open circles). (Adapted from Martin 1988b.)

already been cited to illustrate that the physical environment of the nest may be important in nest site selection, but, comparing different habitat types, it was found that four variables discriminated between nest sites with high and low predation rates. For low predation, the most important of these was the number of small white firs, and the second the extent of cover at the side. Nests experiencing lower predation were also associated with small maple (*Acer*) stems and lower nesting sites, although both these influences were weak. In this species it seems that predation may be a major influence on nest site selection and may be operating at the level of differences in habitat type.

The way in which such site selection behaviour might reduce predation could be explained by imagining that, as the vegetation becomes more dense, nests become more concealed. This *total foliage* hypothesis has been used to explain why increasing numbers of species can co-exist in habitats with more foliage and structural complexity. An alternative hypothesis is that birds select nest sites in areas where there are many potential nest sites, so reducing the probability that a site containing a nest will be discovered by a predator, the *potential prey site hypothesis*. Martin (1993c) tested these hypotheses for site selection on the hermit thrush and MacGillivray's warbler (*Oporornis tolmiei*), and in both cases nest predation was found to be reduced when the nest patch contained more potential nest sites but not when it simply contained more vegetation cover (Fig. 7.4). However, if nest density increases in a favoured patch, this should increase the predation risk to each nest by encouraging greater attention from predators as they experience

Figure 7.4
Daily mortality rate (loss
of nests to predators) in
MacGillivrays warbler
(*Oporornis* (*philadelphia*)
tolmiei) and hermit thrush
(*Catharus guttatus*). Nest
losses are significantly
lower in sites with many
compared to few
potential nest sites, but
not significantly different
between sites with more
or less vegetation cover
measured as total stems
(standard errors shown).
(Adapted from Martin
1993b.)

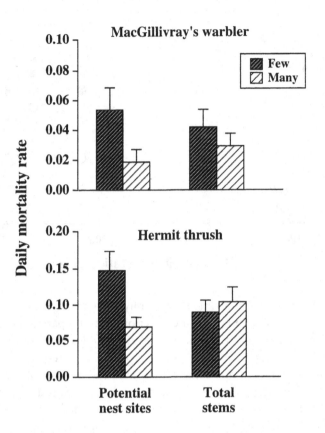

higher reward rates. This was tested by placing artificial basket
nests in small white fir trees, some of which were baited with quail
(*Coturnix*) eggs. Seven nests were placed in each patch and the
number containing eggs varied between patch (one, three and seven
baited nests per patch). Patches with more baited nests were found
to have higher rates of predation (Fig. 7.5, Martin 1988c).

If predation is an important cause of nest failure, then birds with
concealed nests should respond by maximising the distance to that
of their nearest neighbour, producing a uniform (overdispersed)
arrangement. This effect was demonstrated by placing egg-baited
artificial nests in an upland thicket habitat in Canada in three
arrangements, clumped, random and uniform. Lowest predation
rates were experienced by the uniformly dispersed nests (Picman
1988).

Predation pressure also has implications for different species
nesting in the same habitat. Here, selection should favour each
species using a different type of nest site, obliging the predator to
search more substrates, reducing its search efficiency and thus its
overall predation rate. Specialisation in nest site is frequently

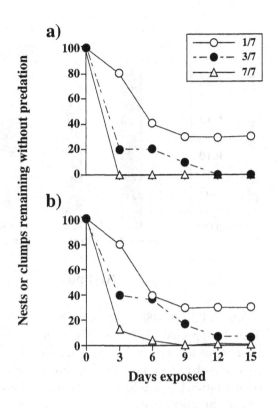

a)

b)

Days exposed

Nests or clumps remaining without predation

1/7
3/7
7/7

Figure 7.5
Predation rates on artificial nests as a function of the ratio of egg-occupied nests to total nests in a clump of seven nests. The three treatments are: one out of seven, three out of seven, and all seven nests per clump containing eggs. (a) Percentage of nest clumps remaining without any nests losing eggs to predators; (b) percentage of intact nests remaining (no egg loss to predators). (Adapted from Martin 1988c.)

observed, but to test whether it had the predicted effect, Martin (1988c, 1993c) used artificial baited nests spread uniformly throughout the habitat but in four different sites chosen to correspond to four local nesting species. These were once again the hermit thrush, with nests placed in small white firs and 1 m from the ground; the orange-crowned warbler (*Vermivora celata*), with nests on the ground under a deciduous shrub; MacGillivray's warbler, with nests placed 1 m up a maple; and, finally, nests placed 3 m above ground in maple to simulate the *black-headed grosbeak (*Pheucticus melanocephalus*). Predation rates on this *partitioned* nest arrangement (i.e. nests in species-specific sites) were compared with those on *unpartitioned* nests (i.e. nests placed at the same density but all in only one of the four nest sites). As predicted, predation rates were higher in the unpartitioned arrangement (Fig. 7.6), supporting the benefit of nest site specialisation.

This so-called *predation/diversity hypothesis* was also tested by Marini (1997), who compared predation of natural translocated nests in 'high nest richness' (nests of five species in typical nest sites) and 'low nest richness' (nests of only two species) sites, controlled for nest density. The sites were located in forested

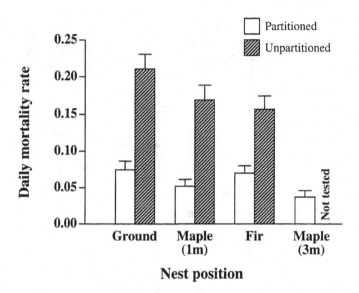

Figure 7.6
Rate at which artificial nests were lost to predators per day when either placed among four different types of site (partitioned: unshaded bars) or placed all in the same type of site (unpartitioned: shaded bars) (standard errors shown). (Adapted from Martin 1993b.)

ravines in Illinois, USA. Despite between-site variability, the results overall supported the hypothesis, with significantly higher predation in the low rather than high nest richness sites. Marini (1997) concludes that natural selection promotes nest site diversity in bird communities, but that local variation in patterns of predation may obscure broader trends.

The importance of predation in shaping nest site selection is not, however, universally accepted. Andrén (1991) criticises Martin (1988c) for using uniformly spaced nests in his experiments rather than comparing predation between a random and a uniform dispersed pattern to show that vulnerability is influenced by nearest neighbour distance rather than just density. Andrén presents data on a natural population of mallard (*Anas platyrhynchos*), showing that nest predation is not influenced by nearest neighbour nest. He also cites studies on nest populations of other species such as the northern lapwing (*Vanellus vanellus*), similarly showing no effect of nearest neighbour nest distance (Galbraith 1988). His conclusion is that predation may be exaggerated as a factor causing spacing out of nests, at least in the low nest densities typical of many species.

The role of predation was, however, demonstrated by Best and Stauffer (1980), who studied the nesting success in a riparian habitat in Idaho in relation to eight nesting variables, some concerned with nest site or general habitat but including nest building date and adult body weight. Predation was a more important factor in nest failure than desertion, brood parasitism or natural disasters.

Predation by birds, snakes and small mammals, was more important than that by large mammals, whose impact diminished in the
higher vegetation layers and on birds of larger size. Predation by
birds, snakes and small mammals collectively was distributed
evenly throughout the vegetation layers, but data on the separate
effects of each of these predators were not shown.

In a study of eight species of pigeons and doves nesting in three
types of forest habitat in Puerto Rico, nesting success was correlated with a variety of attributes of the nest sites. The most important microhabitat variable was found to be nest cover, with nests
constructed on epiphytes losing fewer clutches to predation than
those built on bare branches (Rivera-Milán 1996). However, an
experimental study on nesting success in relation to vegetation
cover of the *hooded warbler (*Wilsonia citrina*) nests within
shrubbery in North American temperate forest failed to demonstrate a correlation between nest concealment and breeding success
(Howlett & Stutchbury 1996).

The complexity of the relationship between nest cover and
breeding success is illustrated in other studies. The scissor-tailed
flycatcher (*Tyrannus forficatus*) nests in mesquite and associated
shrubbery in Texas, where the greatest predation threat is probably posed by snakes. It was predicted that, to aid birds in active
nest defence, the flycatcher nests would be sited with little horizontal cover. The observed horizontal cover at nests appeared to
be a trade-off between greater exposure to counteract predation
and greater cover to reduce the damaging effects of physical factors of wind and sun (Nolte & Fulbright 1996). Cresswell (1997)
found that, while the survival of test eggs placed in *blackbird
(*Turdus merula*) nests after breeding had been completed was
correlated with human ability to detect the nests, this difference
was not apparent for the natural clutches in the same nests when
parents had been in attendance. This suggests that in some way
nest defence was able to compensate for predation risk associated
with a poor nest location.

This draws attention to the fact that nest predators are varied
and that to understand predation threat also requires knowledge of
predator behaviour. An experiment on the effect of environmental
complexity on the ability of racoons (*Procyon lotor*) to locate
artificial nests placed on the ground and containing quail eggs
showed that, in a habitat with the highest level of vegetation variety
and density, racoons increased their total search time and reduced
the number of clutches located compared with study plots with two
lower levels of vegetation complexity (Bowman & Harris 1980).

The degree of concealment was found not to influence nest predation, except where nests were completely in the open. The racoons seemed to be using largely olfactory cues to locate nests. Matessi and Bogliani (1994), placing artificial nests in woodland and woodland-edge sites in Italy, found that well-camouflaged nests were less predated, suggesting in this case that a visually hunting predator was involved. So, whereas the negative relationship between foliage density and predation rates may be generally true (Martin 1992), the ways in which cryptic nests are concealed may vary according to the nature of the predator species.

Fenske-Crawford and Neimi (1997) assessed nest predation using artificial ground nests with associated cameras placed at forest edges defined as either 'hard' or 'soft', depending on the speed of the transition from more mature forest to an open, regenerating one. They found that clutch losses were complicated and that eight different mammalian species were involved. McCown's longspur (*Calcarius mccownii*), nesting on the ground in short grass prairie in Colorado, USA, were found to be more at risk of losing clutches when nests were placed beside shrubs than when they were placed in more exposed sites, apparently because the main nest predator (thirteen-lined groundsquirrels, *Spermophilus tridecemlineatus*) prefers cover when hunting (With 1994). If nest sites are being chosen to protect against more than one species, the optimum choice may vary depending on the predominant local predation pressures. Schaefer (1976) explained the regional variation in the preferred nest site of the *northern oriole (*Icterus galbula*) as due to differences in predominant selection pressures; in an area of high squirrel predation, nests were located on thinner branches compared with an area characterised by high winds.

So, in some habitats siting of the nest may be a compromise between predation risk and other factors affecting breeding success. An example of this is provided by MacGillivray's warbler, which typically nests at a height of 1 m in short maple thicket, but about one-third of nests are located in small white firs where they experience much lower success (Martin 1993c). Part of the explanation for this may be that it allows opportunities for nesting in suboptimal patches for birds which, due to lack of ability to compete for the best sites, would otherwise not breed at all. In addition, it was noticed that, in a year in which leaf development in the maples was unusually late, two-thirds of MacGillivray's warbler nests were located in small firs. So, selection may favour some lack of specialisation in less predictable habitats.

7.5 Predation rates

In complex habitats it is important to establish if certain types of nest site are exposed to greater risk than others. There has been a belief that predation rates were higher for ground-nesting species (Collias & Collias 1984). This was based on an extensive comparative study by Ricklefs (1969) showing that, not specifically predation, but nest failure of whatever kind was greater for ground nesters. More recent studies have, however, shown that the situation is in fact rather more complicated. Martin (1993a), in a study which examined the nesting success of a number of passerine species, found that, whereas in a savannah grassland habitat greater predation losses were indeed experienced by ground nesters compared to shrub nesters, in a forest habitat the ground nesters were least predated (Fig. 7.7). Matessi and Bogliani (1994) similarly found that, in certain woodland sites in Italy, ground nesters were less predated than arboreal nesters. It does, therefore, seem that the relative risk of different nest sites cannot be readily generalised. This does not mean that general principles may not emerge from a more detailed examination of the evidence, as is shown in consideration of the predation risk suffered by cavity nesters.

Nilsson (1986) reported that the breeding success of six natural cavity-nesting species in Sweden did not differ from comparable open-nesting birds. Oniki (1979) found that, in the neotropics, predation rates were significantly lower for cavity nesters compared to cup nesters. Cavity nests were also located in preferred positions; they were seldom at or near ground level or in dark places. Relatively high failure rates were demonstrated for cavity nests located low down or with greater foliage cover in a North American woodland site by Li and Martin (1991). Martin and Li (1992), comparing open and cavity nesters in the same habitat, also confirmed that cavity nests were more successful. However, they found that, if the cavity nesters themselves were divided into *excavators* (e.g. woodpeckers, flickers, sapsuckers) and *non-excavators* (e.g. nuthatches and chickadees), then the former had a greater success than the latter, although the non-excavators alone still did better than open nesters.

The possible reasons for the lower nest success of non-excavators compared with excavators are multiple. It may in part be due to their tendency to nest lower, have more foliage cover and be smaller in body size (Li & Martin 1991). Possibly, in being obliged to re-use a cavity, they are more exposed to ectoparasites (Section 6.7), or to predators that have already learned of the

Figure 7.7
Comparison of nest
predation rates for species
nesting on the ground, in
the shrub layer or in the
canopy shown for three
different habitat types:
forest, shrub/grass and
marsh. (Means and 95%
confidence intervals.)
Ground nesters in forests
experience least predation.
(Adapted from Martin
1993a.)

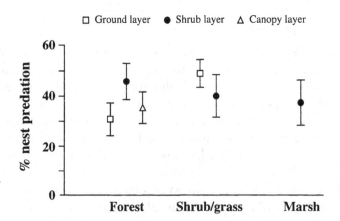

cavity's location; the latter theory is supported by the observation that black woodpeckers (*Dryocopus martius*) have significantly lower success when re-using a nest cavity than when excavating a new one due to differential predation by pine martens (*Martes martes*) (Nilsson, Johnsson & Tjernberg 1991).

The differences in reproductive success of excavators and non-excavators draw attention to the dependence of the latter upon the former. A non-excavator's ability to find a nest site at all may be largely or wholly dependent upon the presence of excavators. A significant effect of excavators on the diversity of species nesting in a woodland habitat is apparent from the compilation of Scott *et al.* (1977), which lists 84 native cavity-nesting species in North American forests. Of these, 26 (31%) nest in natural cavities in trees, 27 species (32%) are nest cavity excavators, and the remaining 31 species (37%) are wholly or at least moderately dependent upon cavities created by excavators. Dobkin *et al.* (1995) found that the abundance of both non-excavators and excavators was positively correlated with numbers of cavities. *Tree swallows (*Tachycineta bicolor*) were heavily dependent upon cavities created by red-naped sapsuckers (*Sphyrapicus nuchalis*), and the mountain bluebirds (*Sialia currucoides*) used predominantly cavities made by *northern flickers (*Colaptes auratus*).The degree of this dependence is illustrated by the sometimes extreme aggression, even resulting in death, shown between excavating woodpeckers and secondary cavity nesters (Section 6.6).

Alerstam and Högstedt (1981) predict two consequences of the competition for nesting cavities. The first is that residents will have priority over migrants, because they have priority of access to the prime sites. The second is that the benefits of acquiring a safe cavity in a very competitive situation may only outweigh the costs for species that feed in exposed niches and so are likely to betray the nest site to visual hunting predators by their visits to the nest. Swedish breeding birds support the predictions that resident species are predominantly cavity nesters (in tree or rock cavities, burrows or houses), and that cavity nesting predominates among exposed site foragers, while unsheltered nests are typical of non-exposed foragers. Nest site limitation and consequent competition should not, however, be assumed to be a universal constraint on non-excavating cavity nests. Brawn and Balda (1988), by providing nest boxes to a forest study site in Arizona, found that only three out of the six secondary cavity nesters increased in population.

7.6 Coloniality and nest defence

Among the passerines, most species are solitary nesters and rather few are colonial (Emlen 1954). A variety of reasons for the evolution of coloniality have been proposed and none seems to provide a general explanation (Wittenberger & Hunt 1985). The dispersed predictable prey exploited by many bird species favours territoriality and therefore spaced nesting, but, for smaller species with consequently small nests, solitary nesting in highly cryptic nests may be favoured as a protection against predation. The larger the bird, the greater the problems of nest concealment, but the better the prospect of active nest defence when nesting colonially. Other aspects of the ecology, however, in particular food dispersion, may still determine whether colonial nesting is possible. This may be true of the colonially nesting *rook (Corvus frugilegus) compared with the solitary nesting carrion crow (Corvus corone), a closely related species of similar size. In the presence of a predator, rooks show collective circling and vocalisation, whereas to the same stimulus a carrion crow tends to fly away from its nest (Röell & Bossema 1982).

Colonial nesting offers two alternative methods of nest protection: the selfish herd effect and group defence. The rook illustrates the latter, but in the former, individuals dilute the risk of predation by association with other nests. Brown cacholotes (Pseudoseisura lophotes), unusually, are able to make use of the decoy effect of other nests even though they are territorial. They make a succession of dormitory nests throughout the year so that, at the time of breeding, a pair may have up to ten nests in their territory. Nores and Nores (1994b) found that the survival of experimental clutches was positively correlated with the number of nests in the territory.

The penalty of predation may be minimised for colonial nesters if nest sites are inaccessible. Nests of the rook seem to be wedged in the forks of the highest suitable branches. The weaverbird *Ploceus cucullatus* nests colonially on inaccessible branches, sometimes over water (Crook 1964). Larger sea-bird species, such as gulls and auks, nest on islands or precipitous cliffs where nest site availability rather than group defence may have led to colonial nesting. However, if collective nest defence is to be claimed as the adaptive explanation of coloniality, then certain predictions should be upheld. These are (1) that colonial nests should be predated less than isolated ones, (2) nests at the colony centre should be safer than those at the periphery, and (3) synchronised nests should be less preyed upon than non-synchronised nests.

Kruuk (1964) demonstrated that the group mobbing response of black-headed gulls (*Larus ridibundus*) reduced predation on experimental nests compared to controls and that, consequently, nests at the edge of the colony were more likely to be predated than colony centre ones. The breeding of black-headed gulls is also highly synchronised and birds breeding outside this peak suffer greater predation levels (Patterson 1965). Andersson and Wiklund (1978) used natural variation in the nesting density of fieldfares (*Turdus pilaris*) to demonstrate that predation was higher in clumped, artificial nests than in spaced ones. However, placing artificial nests close to nesting fieldfares produced a different pattern: predation was lower if such a nest was placed inside or on the edge of a fieldfare colony than when placed near to a solitary nest, because when close to the colony it benefited from the birds' group defence (Fig. 7.8). Why, then, do fieldfares not all nest colonially? The authors suggest that their study, which confined itself to egg predation, may only give a partial answer and that during the chick-rearing phase the balance of advantage may be different.

It must also be borne in mind that within one species the competitive ability of some birds is less than others, so that lower status birds may have to choose the best of a rather limited range of options compared with more dominant ones. Robinson (1986) was able to demonstrate this on colonial nesting yellow-rumped caciques (*Cacicus cela*) in Peru. The most serious predator of this species was the brown capuchin monkey (*Cebus apella*), which was capable of destroying whole colonies. Fledging success within surviving colonies was the result of differential vulnerability to avian predators such as Cuvier's toucan (*Ramphastos cuvieri*) and the black caracara (*Daptrius ater*). The most successful nests were found to be clustered and synchronous compared to clustered and asynchronous (Fig. 7.9a). Isolated nests did least well, with their success depending on other aspects of the nest site such as being

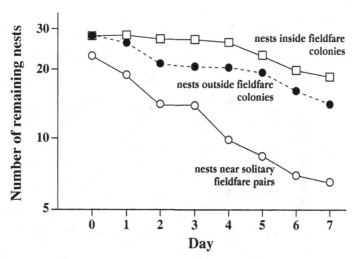

Figure 7.8
Effect of colonial nesting on loss of clutches by fieldfares (*Turdus pilaris*), plotted over seven days. Artificial nests were predated less when placed inside or at the edge of nesting fieldfare colonies rather than near to a solitary nest. (Adapted from Andersson & Wiklund 1978.)

located on an island unreachable by monkeys. Females were observed to engage in aerial fights for possession of favoured sites; by weighing individual birds, it was shown that heavier females were more likely to be clustered and synchronous, whereas the lightest birds were solitary nesters (Fig. 7.9b).

7.7 Sites exploiting the nest defence of other birds

Birds with formidable nest defences (some colonial nesters and some birds of prey) create opportunities for species with a less effective system to profit by nesting in association with them. So black-headed gull colonies may contain smaller groups of common tern (*Sterna hirundo*) (Kruuk 1964). A number of bird species also nest in association with birds of prey (Skutch 1976). These do not of themselves provide adequate evidence either of active association or of benefit. However, evidence for such a *protector species hypothesis* does exist, both for species that associate with colonial nesters and those that associate with birds of prey.

Blanco and Tella (1997) found that red-billed choughs (*Pyrrhocorax pyrrhocorax*) preferred to nest in old buildings containing nesting colonies of lesser kestrel (*Falco naumanni*) than alone. Both species share nest predators in common, but the kestrels were shown to be more vigilant and vigorous nest defenders. The nest failure of choughs associating with kestrels was found to be only 16%, but for those nesting alone in old buildings it was 65%. Tremblay *et al.* (1997) showed that the nesting success of great snow geese (*Anser caerulescens*) was greater when nesting within

Figure 7.9
(a) The fledging success of
yellow-rumped caciques
(*Cacicus cela*) is greater
when they nest
synchronously and in
clusters. Further
protection is provided by
nesting in inaccessible
sites or close to wasp
nests (numbers of nests
shown). (b) Yellow-
rumped cacique females
compete for sites, with
heaviest females showing
the optimum site and
pattern of breeding
(clustered, synchronous
and inaccessible)
(numbers of females
shown). (Adapted from
Robinson 1986.)

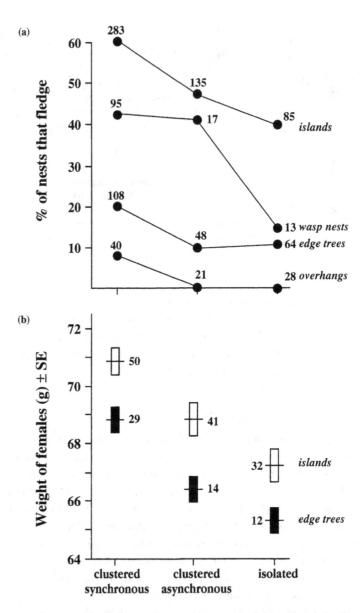

200 m of breeding snowy owls (*Nyctea scandiaca*), from which they derive protection from arctic foxes (*Alopex lagopus*). Bogliani *et al.* (Bogliani, Tiso & Barbieri 1992, Bogliani, Sergio & Tavecchia 1999) showed that, in a poplar (*Populus*) plantation, *woodpigeon (*Columba palumbus*) nests were significantly clustered (within 40 m) at nests of the Eurasian hobby (*Falco subbuteo*) (Fig. 7.10), and that hobbies actively defended this area from predators such as crows. Using egg-baited dummy pigeon nests placed near to, or

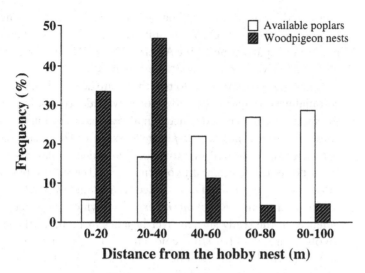

Figure 7.10
*Woodpigeons (*Columba palumbus*) demonstrate a preference for nesting near to the nest of a hobby (*Falco subbuteo*). Graph shows the use of poplar trees at different distances from the nest of a hobby in relation to the number of poplars available. (Adapted from Bogliani *et al.* 1992.)

separate from, a hobby nest they demonstrated that the former survived better from the association and protection was positively correlated with aggressiveness of the hobby pair's own nest defence. However, the association also had costs: hobbies depended upon woodpigeons for 14% by weight of their prey.

Similar studies have been carried out by Ueta (1994, 1998) on an association between the *azure-winged magpie (*Cyanopica cyana*) and the Japanese lesser sparrowhawk (*Accipiter gularis*). He observed a spatial association between nests of the two species, and egg-baited artificial magpie nests were more likely to be predated if not within 50 m of a hawk's nest. The magpies also placed their nests in less concealed sites when they were located close to sparrowhawk nests. However, the hawk's own nest defence behaviour, while active during incubation and nestling phases, dropped sharply at the fledgling stage, resulting in a marked increase in the crow predation on the artificial nests. This emphasises the necessity for the magpies to adapt their nest building not only in space but also in time, to benefit from the hawk. In this case, the benefit to the magpies is not only reduced predation but also abandonment of their own nest defence behaviour. The costs to the magpie are apparently slight as sparrowhawk predation on them is rare. Savannah sparrows (*Passerculus sandwichensis*), studied in Canada, nest in colonies of herring gulls (*Larus argentatus*), a predator on their nests. However, their brood survival was as good in the incubation phase and better in the nestling period than for sparrows nesting outside the colony, because the gulls deter a much more dangerous predator of sparrow nests, the American crow (*Corvus brachyrhynchos*) (Wheelwright, Lawler & Weinstein 1997).

A nesting association exploiting the group defence behaviour of a colonial nesting species provides not only evidence of advantage to the species with no nest defence, but also actual loss of fitness to the colonial species, i.e. *nest defence parasitism*. The association is between sand-coloured nighthawks (*Chordeiles rupestris*), which show no nest defence, and two tern species, the yellow-billed tern (*Sterna superciliaris*) and large-billed tern (*Phaetusa simplex*), and the black skimmer (*Rynchops niger*), all of which show active nest defence. In the study site on river beaches in Peru, all the species involved are found nesting colonially, so it is not immediately apparent which, if any, will benefit by association. For example, it might be that the terns and skimmers benefit by predators concentrating on nests of the non-defending nighthawks. In fact, the opposite proves to be true. Eighty-eight per cent of nighthawk nests successfully hatched eggs when within 10 m of a tern or skimmer nest, but only 54% of them when the distance was greater than 30 m. By contrast, tern and skimmer success was positively correlated with their own numbers, but negatively correlated with the nearby density of nighthawks (Fig. 7.11a, Groom 1992). The presence of large numbers of nighthawks appears to have the effect of diluting the collective defence of the terns and skimmers, so they spend a greater proportion of their time in nest defence and less time in foraging and parental care (Fig. 7.11b), which suggests that there may be little the terns and skimmers can do to avoid this parasitism by nighthawks, which settle on the beach after they have already begun nesting.

7.8 Nest sites associated with arthropods

Members of a variety of passerine families have either facultative or more or less obligate nesting associations with insects or, more occasionally, other arthropods. These associations seem almost invariably to be initiated by the birds and to be for their benefit. However, Keeping (personal communication) records an acacia bush containing several nests of the *white-browed sparrow-weaver (*Ploceopasser mahali*), all of which had a paper wasp (*Polistes*) nest attached to the underside, demonstrating that the wasps chose sites associated with the weavers. The other type of association is one where birds nest inside arthropod nests, possibly to make use of the controlled physical environment they offer, to

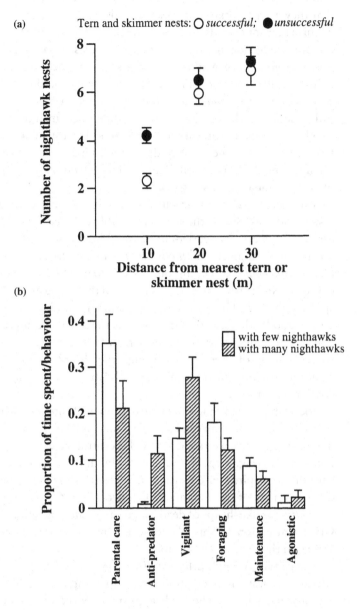

(a)

Tern and skimmer nests: ○ *successful;* ● *unsuccessful*

Number of nighthawk nests

Distance from nearest tern or skimmer nest (m)

(b)

Proportion of time spent/behaviour

□ with few nighthawks
▨ with many nighthawks

Parental care · Anti-predator · Vigilant · Foraging · Maintenance · Agonistic

Figure 7.11
(a) The number of nighthawks (*Chordeiles rupestris*) surrounding the nests of terns and skimmers is shown for three distance categories. The breeding success of terns and skimmers is affected by the number of nighthawks nesting close to them (10 m category) (standard errors shown). (b) When numbers of nesting nighthawks is large, terns and skimmers spend more time in protective behaviours (anti-predator and vigilant) and less time in parental care and foraging (standard errors shown). (Adapted from Groom 1992.)

take advantage of a location which a predator might overlook, or just to exploit a site in which a nest can be built cheaply. Birds in this category are known to nest inside structures built by termites, caterpillars and spiders.

a) Caterpillars and spiders
Very large silk structures relative to their body size are built by some social-living spiders and caterpillars, sufficiently large, in fact,

that they are potential nest sites for birds. A few scattered records confirm that some birds do exploit them. Five out of six records in a museum collection for the *northern beardless-tyrannulet (*Campostoma imerbe*) in Middle America state the location as in old tent-caterpillar tents (Lepidoptera). These tents could be made by species of *Malacosoma* (Lasiocampidae) found in North and Central America (Stehr & Cook 1968). Some other members of this family, including European species, also construct silk tents, as do a few members of other widely distributed families such as Nymphalidae and the 'processional' caterpillars (*Thaumetopoea* species) of Notodontidae (Grassé 1951).

Moreau (1942), summarising African birds' nests recorded from inside spider webs, lists the southern double-collared sunbird (*Nectarinia chalybea*), the scarlet-chested sunbird (*N. senegalensis*) and the amethyst sunbird (*N. amethystina*) and the little grey alseonax (*Muscicapa epulata*). The alseonax is said to nest 'in a great mass of dry leaves and trash that collected in one of the strong and extensive spider webs found in the forest – a customary nesting place'. This gives a good indication that the spider is the social species *Agelena consociata*, found, like the flycatcher, in West Africa. Its colonies weave a mass of sheet webs through the shrubby vegetation to create a structure up to 3 m across (Krafft 1966), well able to contain a bird nest.

Social spiders building similar structures are found in other parts of the tropics. For example in South America *Anelosimus eximius* makes an extensive web structure with a central retreat containing a ball of leaves (Fig. 7.12, Simon 1891). This appears to be a suitable site for small birds to conceal a nest. Sick (1993) enigmatically describes nests of the Tyrannid royal flycatcher (*Onychorhynchus coronatus*) as like 'heaps of dead leaves caught in large spider webs'. Greater attention to this phenomenon might identify more bird species in this type of association.

The gabar goshawk (*Melierax (Micronisus) gabar*) of southern Africa also nests in association with a social spider, in this case a *Stegodyphus* species. The birds appear to collect spiders on their webs and incorporate them into the nest, located 3–12 m above ground, well above the height where the spider colonies normally occur (Henschel, Mendelsohn & Simmons 1991). Nevertheless, in this situation the spiders flourish, apparently on flies attracted to the goshawk nest, until the whole structure becomes festooned in a mass of web. How this benefits the birds is unclear, but the strongest candidate hypotheses appear to be camouflage of the goshawk nests (Steyn 1992) or reduction of arthropod

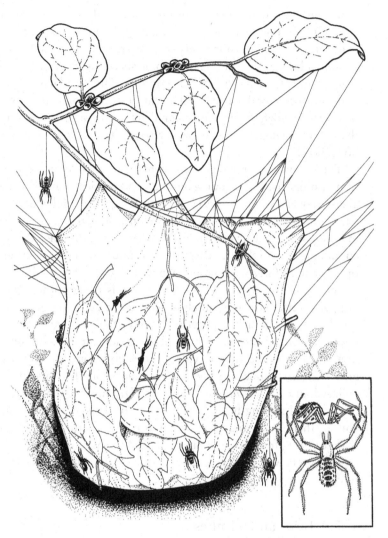

Figure 7.12
The web of the South
American social spider
(*Anelosimus eximius*)
may be a suitable nest site
for some bird species.
(Adapted from Simon
1891, in Wilson 1971.)

ectoparasite attacks on the chicks through spider predation (Henschel *et al.* 1992).

b) Social insects

Termites (order Isoptera) and ants, bees and wasps (order
Hymenoptera) together embrace the most remarkable of all builders excepting only humans. Their nests can be extremely large
relative to the builders themselves and contain huge numbers of
insects. The largest recorded termite mound (*Amitermes* species),
at 6.7 m high (*New Scientist*, 1988), probably contained several
million individuals. Ant colonies also contain up to five to eight

million adults (Hölldobler & Wilson 1990), while those of bees and wasps, more modestly, can attain numbers of more than 150 000 (Michener 1974, Ito 1993). Very large nests like these are built to contain adults, rear brood and, in some species to store food. This makes them attractive targets for certain predators and has resulted in the evolution of defensive nest architecture and often very vigorous defensive behaviour, which can deter predators 100 000 times the size of a single insect. Consider yourself against a handful of honeybees!

These powerful area defence systems are available for exploitation by other species including birds, and, in tropical habitats where a nest may last several years (Jeanne 1975), quite complex dependencies and interdependencies may develop around them. I have seen in Costa Rica an acacia thorn bush with a nest of the wasp *Polybia* containing several thousand adults, surrounded by a dozen or more open comb *Polistes* wasp nests, each with a score or so adults, near to which are roosting large-bodied Katidids (long-horned grasshoppers, Tettigoniidae). Stingless bees of the genus *Trigona* may build nests inside arboreal termite nests (*Nasutitermes*) and ant nests (*Azteca*) (Michener 1974). This relationship may be one of mutual defence, the ants providing protection from raids by army ants (*Eciton*), while the bees deter attacks on the ants by the collared anteater (*Tamandua tetradactyla*) (Smith 1980). Finally, like colonial nesting birds, social insects of the same species may cluster their nests together, apparently for mutual protection. Jeanne (1978) found that in Brazil 92% of the paper wasp (*Polybia rejecta*) nests were in aggregations, one an impressive cluster of 23 nests.

7.9 Birds and termites

Primitive termites live inside and feed upon dead timber; others live in the ground or in mounds projecting above the ground. The material of these mounds is predominantly soil mixed with salivary or faecal cement which forms a hard outer layer. The mounds, notably of *Macrotermes* and *Amitermes* of Africa, Asia and Australia, may be 2–6 m high and so offer similar nesting potential to birds as an earth bank or tree trunk. In addition, both these genera have highly developed systems for maintaining relatively constant internal nest temperatures (Luscher 1961, Grigg 1973).

There are also arboreal termites of the genus *Nasutitermes* found in the New and Old World, which build massive globular nests up to 1 m across. These nests are made of partially rotted wood mixed with faecal cement (carton). Their thin outer wall and cellular structure make them about as easy for a bird to excavate as partially rotted wood.

Some termite species have soldiers with large jaws and a powerful bite, but, as termites abhor the light, colony defence generally does not extend beyond the nest surface. For birds, therefore, the attraction of their nests seems to be that they are large objects inside which cavities can be excavated, or possibly that the termites maintain a regulated nest environment. The association of a number of bird species with termite nests is facultative, but in spite of this no studies have compared their breeding success in termite nests with that in tree or subterranean cavities and no experimental studies have been undertaken.

The number of species of birds adopting termite mound nest sites is not large; the nest survey reveals 71 species across nine families. They are characterised by virtually all belonging to non-passerine families and being in the size range 40–80 g; that is, neither large nor small for birds. One group that is particularly dependent upon these sites comprises the three families collectively referred to as kingfishers (Alcedinidae, Dacelonidae and Cerylidae), although an equally large number is found among the parrots (Psittacidae) (Table 7.1). Some members of all these families excavate their own cavities. Across all these families, nesting is predominantly in arboreal termite nests.

Methods of excavation of nests in termitaria are derived from methods used by that species or its near relatives to deal with soil or wood, and vary with beak type. Hindwood (1959) rather dramatically reports kingfishers commencing to excavate termitaria by flying at them 'from a distance of several feet or more', striking the nest with their pointed beaks. By contrast, Hardy (1963) describes the orange-fronted parakeet (*Aratinga canicularis*) excavating with biting and chewing movements which involve so little body movement that from a distance the bird appears to be doing nothing.

7.10 Birds and ants

The approximately 9000 ant species are all social and belong to a single family (Formicidae) of the order Hymenoptera. There are many ground-dwelling ant species, but none makes large mounds like termites. For birds, the important ant species live in the trees. In the New World, ants of the genus *Azteca*, and in the Old World, *Crematogaster*, make large carton nests which may be confused with those of arboreal termites. In addition, there are ants that

Table 7.1. **Records of birds nesting in terrestrial termite mounds or arboreal termite nests**

	Total	Nesting in termite nests	Percentage of family	Bird weight (g)	In arboreal (A) or terrestrial (T) termite nests
Picidae					
S	215	3	1.4	39–69	A +
G	28	2	7.1	(n = 2)	
Galbulidae					
S	18	2	11.1	16.7	A + and T +
G	5	1	20.0	(n = 1)	
Bucconidae					
S	33	3	9.1	59–69	A +
G	10	3	30.0	(n = 2)	
Trogonidae					
S	39	6	15.4	73–141	A +
G	6	1	16.7	(n = 4)	
Alcedinidae					
S	24	4	16.7	11–27	A +
G	3	3	100	(n = 31)	
Dacelonidae					
S	61	21	34.4	35–311	A + +
G	12	8	66.6	(n = 16)	
Cerylidae					
S	9	2	22.2	15.8	A +
G	3	1	33.3	(n = 1)	
Psittacidae					
S	358	31	8.4	11–510	A ++++ and T +
G	80	14	17.5	(n = 20)	

All eight families are non-passerine. The bird weights are shown as the range for species recorded in Dunning (1993); n = smple size; S = species; G = genera; + symbols denote the degree to which the association is obligate.

Information for this table was obtained from Campbell (1901); Coates (1985); Forshaw (1989); Hindwood (1959); Moreau (1942); Myers (1935); Sick (1993); Skutch (1976); Weaver (1982).

make small purse-like nests of leaves held together with larval silk. Most important of these for birds are weaver ants (*Oecophylla*), whose two species of moderate-sized, aggressive, red ants range from Africa through Asia to Australia. Each individual *Oecophylla* nest is only about 20 cm long and contains only a few hundred workers, but one colony can consist of hundreds of nests and defend a territory occupying several mature forest trees (Hölldobler & Wilson 1990).

The third category of arboreal ant of importance to birds embraces ants that live inside cavities provided by certain species of tree or bush. In return, these plants benefit from the ants' control of insect or vertebrate defoliators. The best studied example of this in the context of bird nest site selection is the New World ant genus *Pseudomyrmex*, which lives inside hollow thorns of some acacia species. Early reports of birds nesting inside ant nests (e.g. Campbell 1901) are unreliable because of confusion between arboreal ant and termite nests. However, the rufous woodpecker (*Celesus brachyurus*) and buff-spotted woodpecker (*Campethera nivosa*) are reported to nest inside nests of *Crematogaster* (Hindwood 1959).

Evidence of protection of the clutch for birds nesting in the proximity of arboreal ant nests is not simply spatial association, but statistical support for active choice by the birds, which results in enhanced breeding success. The eastern chanting-goshawk (*Melierax poliopterus*, Accipitridae) and the red-faced crimson-wing (*Cryptospiza richenovii*, Passeridae) are both known to nest in association with weaver ants (Moreau 1942). The bronze munia (*Lonchura cucullata*, Passeridae) was shown by Maclaren (1950) to be more likely to nest in trees containing weaver ant colonies than in those that did not. These associations are not only tropical. Haemig (1999) demonstrated that tits (*Parus* species) nesting in the interior of a Swedish forest avoided nesting in trees frequented by wood ants (*Formica aquilonia*); however, at the forest edge, where nest predation was much higher, they showed a significant preference for nesting in ant-associated trees.

A small number of assorted bird species nest in association with the ant-acacia thorn bushes of the dry forests of the neotropics, the thorns of which are colonised by *Pseudomyrmex* species. These include the *great kiskadee (*Pitangus sulphuratus*, Tyrannidae), the barred antshrike (*Thamnophilus doliatus*, Thamnophilidae), yellow-billed cacique (*Amblycercus holosericeus*, Fringillidae) and some oriole species (*Icterus*, Fringillidae) (Janzen 1969). Young, Kaspari and Martin (1990) examined whether four species of bird nesting in the ant-acacia (*Acacia collinsii*) were actively selecting trees which gave better ant protection. The birds were the rufous-naped wren (*Campylorhynchus rufinucha*, Certhiidae), the *streak-backed oriole (*Icterus pustulatus*, Fringillidae), and two Tyrannidae, the *yellow-olive flycatcher (*Tolmoyias sulphurescens*) and the great kiskadee.

Figure 7.13
The intensity of the activity of two *Pseudomyrmex* ant species: (a) *P. spinicola* and (b) *P. flavicornis* before and after disturbance of the tree. Rufous-naped wrens (*Campylorhynchus rufinucha*) chose trees with the most vigorous ant response compared to trees chosen by other bird species or with no bird nests at all (standard errors shown). (Adapted from Young *et al.* 1990.)

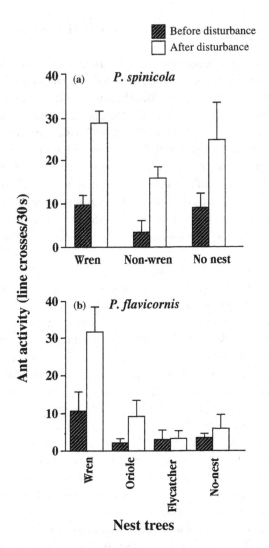

Observations on the ants showed that the two *Pseudomyrmex* species responded more vigorously to disturbance of the acacia tree than non-stinging ant species; also, *Pseudomyrmex spinicola* was more vigorous in defence of its colonies than *P. flavicornis*. The flycatcher, kiskadee and oriole were found to select trees at random with respect to ant species, but the rufous-naped wren nested in trees containing *P. spinicola* twice as often as expected by the relative frequency of the ant colonies, and when selecting a tree occupied by *P. flavicornis*, chose one with a more vigorous than average ant colony (Fig. 7.13). To test whether acacia bushes provided better nest protection than non-acacias, artificial egg-baited nests were placed in both sites. Contrary to expectation, higher predation was experienced by nests in acacias. However, the majority of eggs in acacia trees were damaged by a single beak puncture mark, which Young *et al.* (1990) interpret as made by rufous-naped wrens trying to deter rivals from preferred nest sites.

7.11 Birds and bees

Only one bee family contains highly social species, the Apidae. It includes the bumblebees (Bombini), whose rather small, weakly defended colonies attract no bird nesters, the honeybees (Apini) and stingless bees (Meliponini), with which only a few bird species are known to associate in spite of their large colony sizes. One problem for the birds may be that social bees are essentially cavity nesters, typically in hollow tree trunks or limbs, which may not have suitable bird nest sites close to them. Honeybees (*Apis* species), despite their powerful stings and large colony sizes, are apparently of little importance to birds, although Myers (1935) reports that the yellow weaver (*Ploceus megarhynchus*) nests in association with the honeybee (*Apis indica*). Of slightly more importance to birds are stingless bees which, despite their name, can be extremely aggressive and have a powerful bite which, in some *Trigona* species, is venomous and raises painful blisters on humans (Michener 1974). The *chestnut-headed oropendola (*Psarocolius wagleri*) and the yellow-rumped cacique (*Cacicus cela*) are found near to *Trigona* nests, although they are more commonly associated with those of wasps. Smith (1980) found that one of the most serious causes of mortality for the chicks of these species was larvae of botflies (*Philornis*) burrowing through their skin and into their bodies. Nests placed close to wasp nests, but more particularly *Trigona* colonies, seemed to gain protection from intruding parasitic flies.

7.12 Birds and wasps

The most common association between birds and social Hymenoptera is with wasps. The biology and colony sizes of the social wasps are somewhat variable. The advanced social wasps occur in two subfamilies of the Vespidae, the Vespinae (yellowjackets and hornets) and the Polistinae (paper wasps). The Vespinae are generally northern temperate in their distribution and found colonies with a single queen in the spring. Bird associations with them are unknown, probably because the wasp colonies are still small at the time the birds are breeding. It is the Polistinae, with their predominantly tropical and sub-tropical distribution, with which birds form nesting associations. Some genera, like *Polistes*, are

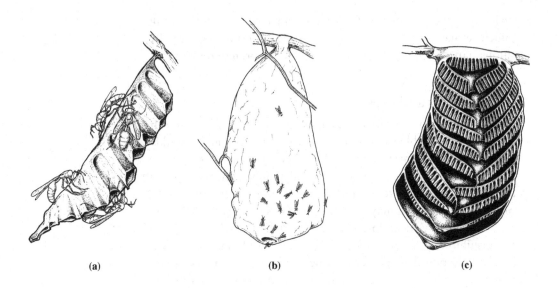

| (a) | (b) | (c) |

Figure 7.14
Nests of tropical social wasps vary greatly in size. The open-comb nest of a *Ropalidia* species (a) is generally attended by fewer than 100 adults, whereas large colonies of a *Chartergus* species – (b) external and (c) internal view – may contain several thousand.

founded by a single or small number of females and generally have colonies of less than 100 adults on single, open-comb nests (Fig. 7.14), e.g. *Belanogater*, *Ropalidia* and *Mischocyttarus*. The majority of genera of Polistinae are in larger colonies, generally swarm founded. Their greatest diversity is in South America and includes colonies of tens of thousands or more adults in durable enclosed nests with a small entrance hole (e.g. *Polybia* and *Chartergus*). These polistines vary from species to species in the vigour of their nest defence, but all are equipped with painful stings.

The number of species of birds nesting in association with wasp nests is quite small; a search of the literature revealed 28 species (Table 7.2). The birds are scattered among nine families, none of which can be said to specialise in the association. There are, in fact, no very strong features they bear in common, although they are generally, but not universally, quite small birds (less than 20 g) and all but one are passerines. Only one bird species is known to nest actually inside a wasp nest, the violaceus trogon (*Trogon violaceus*), which Skutch (1976) observed systematically eating the wasps before taking over the defenceless structure. In all other recorded examples the birds nest near to active wasp colonies to take advantage of their hostile responses to intruders.

The identity of the wasps in these associations is often unrecorded and insufficient is known to determine to what extent the relationship is simply occasional and facultative. For some species, however, the presence of a wasp nest seems more or less essential

Table 7.2. Records of birds nesting in association with wasp nests, together with the genus of the wasp where recorded

Family		Number of associating species out of total	Bird weight (g)	Associated wasp genera
Sibley & Monroe (1990)	Howard & Moore (1991)			
Non-passerine families				
Trogonidae	Trogonidae	1 (39)	64	
Passerine families				
Tyrannidae	Tyrannidae	4 (537)	12–61	*Polybia*
				Synoeca
Furnariidae	Furnariidae	1 (280)	14.8	
Pardalotidae	Acanthizidae	4 (68)	6–8	
Corvidae	Platysteiridae	1 (647)	13	*Polybioides*
Certhiidae	Troglodytidae	1 (97)	30	*Polybia*
Nectariniidae	Nectariniidae	2 (169)	7–10	*Polybioides*
				Ropalidia
Passeridae		8 (386)		
	Estrilididae		9–19 ($n = 5$)	$\Big\{$ *Belonogaster*
	Ploceidae		17–34 ($n = 3$)	*Polybiodes*
				Ropalidia
Fringillidae		6 (993)		
	Emberizidae		9–13 ($n = 2$)	*Polistes*
	Coerebidae		9 ($n = 1$)	*Polybia*
				$\Big\{$ *Polistes*
	Icteridae		78–214	*Polybia*
			($n = 3$)	*Protopolybia*
				Stelopolybia

Bird weights are shown as the range for which weights are available in Dunning (1993). Information for this table was obtained from Chapin (1954); Chisholm (1952); Contino (1968); Haverschmidt (1957); Hindwood (1959); Joyce (1993); McCrae & Walsh (1974); Moreau (1942); Myers (1935); Poulton (1931); Sick (1993); Smith (1968b); Wunderle & Pollock (1985).

for breeding e.g. the *red-cheeked cordonbleu (*Uraeginthus bengalus*, Passeridae), which has been found close to *Belonogaster griseus* (Poulton 1931) and *Ropalidia cincta* colonies (McCrae & Walsh 1974). For the dull-coloured grassquit (*Tiaris obscura*, Fringillidae), there is a statistical demonstration of association with *Polistes canadensis* (Contino 1968). The bananaquit (*Coereba flaveola*, Fringillidae) was shown by Wunderle and Pollock (1985) to prefer trees containing *Polybia occidentalis* and generally to site

Figure 7.15
Nesting attempts by
rufous-naped wrens
(*Campylorhynchus
rufinucha*) were more
successful in terms of
fledging young when a
wasp nest was present
rather than absent in the
two years of study (Joyce
1993).

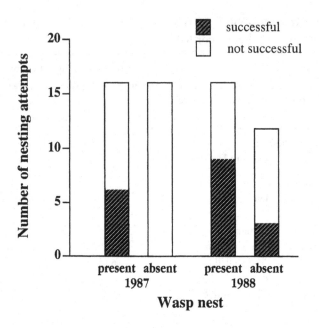

Figure 7.15
Nesting attempts by
rufous-naped wrens
(*Campylorhynchus
rufinucha*) were more
successful in terms of
fledging young when a
wasp nest was present
rather than absent in the
two years of study (Joyce
1993).

the nest within 1 m of them. Nests in this position were significantly less likely to be predated than ones without a wasp nest, and a male bananaquit was more likely to retain a female during the breeding season if he had a wasp nest in his territory.

Further evidence of advantage gained by this behaviour comes from studies on the rufous-naped wren, with its nests located in ant-acacia thorn bushes, because these nests may also be in close proximity to a wasp nest (Fig. 7.15). Joyce (1993) located a number of these wren nests and created from them two comparable groups of nests; in one, each wren nest had an active colony of the highly aggressive wasp *Polybia rejecta* beside it, while those in the other group did not. The wasp nests were acquired by the somewhat hazardous method of collecting each one in a bag at night, a distance of 5–10 km from the study site, then securing it in a thorny, ant-infested tree next to a wren nest. In this way it was demonstrated that wrens provided with a wasp nest had a significantly greater success in fledging the young than wrens without one. Also, in 15 cases in which nesting was attempted in the same tree before and after a *Polybia* nest was introduced, fledging success was greater after the introduction. The protection the wasps provide seems to be particularly against *Cebus* monkeys, that were seen to be repelled by wasp attacks.

The ferocity of the wasps raises the question of whether the birds themselves face risks by nesting so close to them. The evidence here is circumstantial, but in all suggests that the birds do face risks from both wasps and stingless bees, and have evolved some adaptations to minimise them. Smith (1980) observed that oropendolas and caciques attempting to nest near colonies of wasps or stingless bees were quite commonly attacked and, in response to *Trigona* bee attacks in particular, fled until the bees calmed down. Also, whereas adult birds, if stung by the wasp *Protopolybia pumila*, seemed unperturbed, chicks would die. By the time the birds have completed their nest, the bees or wasps have become generally unresponsive, although they will still attack and successfully drive off nest predators such as toucans (Smith 1980). It seems the insects become habituated to the movements of the resident birds while still roused by a novel stimulus.

Smith (1980) also suggests that at least some of this category of birds may be protected from wasps and bees by a deterrent odour, which in some species of adult oropendolas and caciques is very obvious to humans. There is no evidence to support this theory, but it does accord with the supposed deterrent function of the strong, unpleasant smell of the red-throated caracara (*Daptrius americanus*), which is a specialist predator of neotropical wasps and stingless bees (Meliponini).

The vulnerability of the chicks may be reduced in another way – the design of the nest, which is characteristically domed in these species. There are other explanations for nests being roofed (Section 5.5b), so this is rather a weak argument, but support for the wasp attack hypothesis comes from the nest design of the dull-coloured grassquit (*Tiaris obscura*), which nests in close association with nests of *Polistes canadensis* (Contino 1968, Skutch 1976). The nest of *T. obscura* has a cup depth of 200 mm, a thick wall of compactly woven grass flowers, and weighs about 25 g, whereas nests of 11 species of the arguably congeneric seedeaters (*Sporophila*) (Sibley & Monroe 1990), none of which associates with bees or wasps, have nests of about 45 mm depth which are so thin walled as to be transparent and weigh only about 2 g (Contino 1968).

CHAPTER 8

Bowers, building quality and mate assessment

8.1 Introduction

Bowers are structures built by a mere 17 or 18 species of birds belonging to only a single family (Ptilonorynchidae) confined to New Guinea and parts of Australia (Gilliard 1969, Cooper & Forshaw 1977, Sibley & Monroe 1990). These structures would hardly seem candidates for dominating a whole chapter, but for two things. The first is the complexity and to our eyes even beauty of the structures built by some species. Diamond (1987) describes the structure built by the vogelkop bowerbird (*Amblyornis inornatus*) as 'the most elaborately decorated structure erected by any animal other than humans'. The second is that, as Darwin proposed in *The Descent of Man* (1871), the evolution of the elaborate structure of these bowers is the consequence of sexual selection through female choice.

Bowers are not nests; they are built only by males of this family for the purpose of attracting mates. Females alone build nests, in quite separate locations. Debate and active experimentation surround the more general question of female assessment of male quality and the effect this will have on male characters through sexual selection (Andersson 1994). One purpose of this chapter is to show how studies on bower construction, and to a lesser extent on nest building, are making a significant contribution to the understanding of sexual selection through female choice. The other is to examine how bowers actually function to enhance male mating success.

In the majority of birds, males contribute in some way to parental care. In such cases we would expect females to assess, before mating, those characters that might reflect a male's ability to perform parental duties; for example, the courtship feeding of sandeels (*Ammodytes* spp.) by male common terns (*Sterna hirundo*) may be used by females to assess the ability of males subsequently to feed the young (Nisbet 1977). Nest building also is a parental role. It is often performed wholly or largely by males (e.g. weaver birds, Ploceinae), although more commonly by both sexes or the female alone (Collias & Collias 1984). Where the male alone builds, a female might assess male quality on the basis of his nest before deciding whether to mate or not; where both sexes are involved, both partners could use assessments of the other's nest building to determine their level of reproductive investment (Soler, Møller & Soler 1998).

In species in which males take no part in parental care, they are frequently characterised by elaborate and colourful plumage, strikingly different from that of females, suggesting that it is the consequence of sexual selection through female choice. In the bowerbirds also, the males are to a greater or lesser extent brighter in their plumage than females, but the characteristic difference is the use of a bower by the male as a display. In 11 bird families males display in groups, or *leks* (Höglund 1989), allowing simultaneous comparison by females of several males. In some species the displaying males are somewhat separated from one another in a dispersed lek, but still capable of being visited and compared by a female in the days preceding mating. The distribution of the bowers of bowerbirds is, at least in some species, somewhat aggregated, roughly corresponding to a dispersed lek, but if female bowerbirds are choosing males at least partly by a comparison of the properties of their bowers, what criteria are they using and why?

8.2 Sexual selection

The field of sexual selection and mate choice is currently one of active research and theoretical development (Andersson 1994). Two hypotheses for the evolution of exaggerated male displays through female choice, *Fisherian runaway* and *good genes*, have now been joined by other interpretations (Borgia 1986, Ryan 1997).

The Fisherian runaway hypothesis bears the name of its author (Fisher 1930), who stated that a quite arbitrary male character (for example colour of tail feathers), if it becomes preferred by some females, could spread through the male population simply because of that preference and become more and more exaggerated until checked by some counter-selection. Although subsequently criticised on the grounds of circularity, Kirkpatrick (1982), amongst others, using quantitative genetic models, has shown that this sort of process could result in the evolution of exaggerated traits. It has proved difficult to confirm this hypothesis. This may in part be because the evolution can take place very rapidly and so is unlikely to be observed (Ryan 1997), or because it may only be provisionally accepted in the absence of empirical evidence supporting a good genes interpretation.

The good genes hypothesis predicts that female choice is not arbitrary, but ensures that females produce offspring better able to survive and therefore reproduce. For example, Halliday (1978) suggests that choosing an older male is selecting one with a proven capacity to survive. Petrie (1994) was able to show that the survival of peafowl (*Pavo cristatus*) in a near-natural environment was associated with the quality of the eyespot display of the father's tail, so females choosing more showy males produce better quality offspring.

Females might judge the quality of a male through the competitive ability he shows in a lek environment. Competition for favoured sites in the lek could act as a *marker* of male quality. Alternatively, as suggested by the *handicap* hypothesis (Zahavi 1975), elaboration of display ornaments could demonstrate a greater genetic quality for survival in spite of a handicap. To overcome some of the original objections, this hypothesis can be expressed as a condition-dependent handicap: i.e. that male investment in the handicap will depend on his condition in order to optimise the trade-off between mate attraction and survivorship (Ryan 1997). To support this hypothesis, the display needs to be costly, possibly also reducing survival (Lande 1981).

To Fisher's runaway and good genes hypotheses need to be added two more: the *passive choice* and the *proximate benefit* hypotheses. The passive choice hypothesis (Parker 1983) maintains that elaborate displays might evolve simply if such males were more readily detected by females than more modest ones, resulting in more conspicuous males producing more offspring. The proximate benefit hypothesis proposes that males are chosen on the basis of some direct, immediate benefit they offer to females.

The nature of the proximate benefits might vary according to species, but among those suggested are that the quality of the male display could indicate his state of health and therefore the likelihood that he will transmit to the female venereal diseases or ectoparasites (Ryan 1997), or provide a sufficient quantity of viable sperm (Borgia 1986). A related direct selection hypothesis has been referred to as *sensory exploitation*. This envisages that the female selection process might be a pleiotropic effect of selection for sensory capacities that function in other contexts such as prey detection (Ryan 1997). However, although this might provide an initial criterion for female choice, males varying in that character should then be exposed to sexual selection.

Bowers, unlike display plumage, can be experimentally manipulated without touching the male; thus, they offer excellent models for testing these competing explanations of female choice. The last 15 or more years have seen a great increase in our knowledge of the appearance of bowers and the behaviour of their associated males, together with good experimental studies on the nature of female choice. This has helped clarify and develop our understanding of sexual selection.

8.3 Nest building and sexual selection

Nest building is an integral part of reproduction for the majority of birds and in a number of species is clearly integrated into courtship behaviour (Collias & Collias 1984), with ritualised displays involving nest materials held in the beak. For example, in the blue-footed booby (*Sula nebouxii*), nest building is reduced to the exaggerated presentation of tiny pieces of vegetation by the male to the female (Nelson 1978), a trait that might well be expected to be under the influence of sexual selection (Darwin 1871). So, although Chapters 2 and 5 clearly demonstrate the role of natural selection in shaping features of the nest, this does not exclude the possibility of nest features also being under the influence of sexual selection. There is now evidence that, in some cases where males alone build all or the initial part of the nest, the quality of male performance may affect female mating decisions, and that, where both partners build the nest, nest quality affects individual reproductive investment.

A male *village weaver (*Ploceus cucullatus*) hangs from his newly built nest in his yellow and black plumage, flapping his wings and calling. An interested female, on arrival, inspects the nest for as much as 20 minutes and, if satisfied, will line the nest with soft grass and feathers before mating takes place (Collias & Victoria 1978). Nests are built with strips of fresh green grass, but these soon turn brown and become brittle. Experimental manipulation of nests shows that females prefer freshly built nests. Colour and external appearances are not as important to females as strength of the fabric, although males tear down rejected brown-painted nests more than three times as often as rejected green-painted ones (Table 8.1, Collias & Victoria 1978, Jacobs, Collias & Fujimoto 1978).

In a few species, males build several nests in one season and more than one may be shown to a prospecting female. In the *winter wren (*Troglodytes troglodytes*), a male generally builds five or six, and up to ten to 12 nests per season (Garson 1980). Males with a large number of vacant nests are more likely to attract a female to

Table 8.1. **Rejected nests torn down by males before nests reached eight days of age**

Colour	Number of rejected nests*	Rejected nests torn down before eight days old	Percentage of rejected nests torn down by male	p (z-test)
Painted green	37	6	16.2	0.003
Painted brown	35	30	56.6	

* Those nests not permanently accepted by a female.
Male village weavers are more likely to tear down rejected nests dyed brown than those dyed green in order to build a new nest to attract a female.
Adapted from Collias and Victoria (1978).

Figure 8.1
The positive relationship between the number of nests built by territorial male winter wrens (*Troglodytes troglodytes*) and the number of females making breeding attempts on their territory. (The larger the symbol, the more observations at that point.) (Adapted from Evans & Burn 1996.)

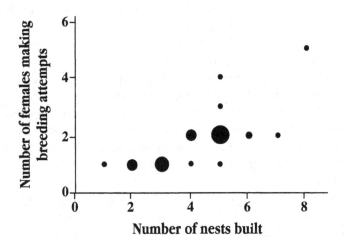

settle than males with fewer nests (Fig. 8.1, Evans & Burn 1996). Males do tend to increase the number of nests they build with age, but the large variation between individuals suggests that some males are just better builders than others and that improvement with age does not compensate for lack of ability. Female *winter wrens do therefore appear to be using nest building as an indicator of some inherent male quality, although it is not clear how this may benefit them.

In long-billed marshwrens (*Telmatodytes palustris*), the male attracts a female into his territory by singing and then shows her a number of nests which she then carefully inspects (Verner 1964). The mean number of nests built by males is about 23 and there is a significant positive correlation between male mating success and number of nest sites (Verner & Engelsen 1970). In neither this species nor the *winter wren is there evidence that quality, as distinct from quantity, of building influences female choice. However, Hoi, Schleicher and Valera (1994) found that not only were female *Eurasian penduline tits (*Remiz pendulinus*) more likely to choose mates that built a larger nest, but also that nests of larger size provided better insulation.

Nests of the magpie (*Pica pica*) are built by both sexes, although the larger twigs and mud for the nest cup are brought predominantly by the male. Nests are variable in size, the larger ones being characterised by a dome of twigs over the nest cup, which has been explained as a protection against predators such as crows (Birkhead 1991). However, Soler *et al.* (1995) found no relationship between nest size and protection against predation. They suggest instead that, since larger nests are associated with better quality territories, this could be a signal by the male of his quality and willingness to invest in reproduction. An effect of male nest building quality on female investment is shown by the black wheatear (*Oenanthe leucura*). In this species, mated pairs carry large numbers of stones to place in nest cavities or other sites. An average of 1–2 kg of stones is carried largely by the male. Moreno *et al.* (1994) found that, when a greater number of stones was carried, females laid earlier, laid larger clutches and fledged more young.

Soler *et al.* (1998) tested the generality of the influence of sexual selection on nest building in species in which both sexes contributed, by making the following predictions: (1) that nests built by both sexes should be bigger than those built by females alone; (2) that if nest building effort is an indication of willingness to invest in parental care, then there should be an increase in parental care for species that build together compared with species in which the female builds alone (because each can assess the other more reliably than in species in which only one sex builds); (3) if 1 and 2 are true, then species that build a larger nest in relation to body size should invest more in reproduction. Evidence obtained from 76 passerine species confirmed the first two predictions and gave some support for the third.

8.4 Court displays and male quality

In springtime at traditional sites in northern Europe, male black grouse (*Tetrao tetrix*) gather at dawn to posture and croon in the presence of choosy females. Similar lek displays in relatively open sites are shown by other species, for example by the ruff (*Philomachus pugnax*) and, in North America, by the *sage grouse (*Centrocercus urophasianus*). These leks have the characteristics that male distribution is clumped, females are freely able to make choices between them, males provide no parental care, and the display sites have no resources of value other than sperm (Bradbury 1981). Some species build a distinctive court or bower as the display site. For these, evidence is either insufficient or unclear on whether the term lek or, for less clumped displays, 'exploded' or dispersed lek is the more appropriate.

In forest habitats, displaying males may be restricted in their movements and difficult for females to compare unless a special display area is prepared. In the *Guianan cock-of-the-rock (*Rupicola rupicola*) (Gilliard 1962), and white-bearded manakin (*Manacus manacus*) (Snow 1962, Lill 1974), males clear patches of forest floor in the area where they collectively display. The male superb lyrebird (*Menura novaehollandiae*) clears an area of forest floor about 1.5 m across and creates in the middle of it a mound of earth 10 cm or 15 cm high on which to stand when displaying his elaborate plumage (Smith 1968a).

A male great argus pheasant (*Argusianus argus*) prepares an individual court on the forest floor of up to 72 m² with vigorous wing beats to clear dead leaves and energetic throwing with the beak to remove sticks, although maintenance of the arena only takes 0.5% of the bird's time (Davidson 1981, 1982). These preparations probably help to make the male's magnificent display of occelated feathers more impressive to visiting females, but there is no evidence that the court itself forms part of the display. This may also be the case in the displays of that strange and vanishing New Zealand ground parrot the kakapo (*Strigops hapbroptilus*). Males in this species apparently display in a dispersed lek by immensely inflating their chests to make deep, booming noises (Merton, Morris & Atkinson 1984). To direct the sound, males boom into bowl-shaped depressions, part natural and part excavated. Each male maintains a 'bowl system' comprising six to ten bowls connected by well-worn paths which may be 50 m or so in extent. The careful clearing and preparation of the soil in each bowl are consistent simply with the improvement of its sound-reflecting properties; however, Merton et al. (1984) also record that the tracks themselves may be meticulously cleared of debris and the flanking vegetation clipped well back. This raises the question of whether the bowl system itself could be part of the male's display when the female actually reaches him.

For Jackson's widowbird (*Euplectes jacksoni*) experimental evidence supports this interpretation. The species lives in African savannah grassland where males gather in leks, each individual preparing its own display court. To attract females, a male leaps up above the tall grass to display his dark plumage and flamboyant tail feathers. Experimental alteration of these feathers has shown that males with shortened tails receive fewer female visits than do unaltered controls and that female visits are correlated with male reproductive success (Andersson 1989, 1992). However, the display court itself is not simply an area of flattened grass, for at its centre is an oval tuft with two shallow recesses, one on each of its long sides, shaped and trimmed with the male's beak. When a female arrives at the court, attracted by the jumping male, she often sits facing this tuft, pecking at the recess while the male hides on the other side. This may last a minute or so before the male appears and makes a copulation attempt. This suggests that the tuft and the recesses form part of the male's display.

Andersson (1991) experimentally altered the central tuft, in some cases ruffling it to make the recesses shallower and less distinct (*impaired*), in others using his scissors to make the recesses deeper and more distinct (*improved*). He found that females visiting impaired courts stayed too short a time for males to succeed in approaching them, whereas those arriving at improved courts stayed longer than at normal courts. Indeed, all copulations observed occurred either at normal or improved courts (Fig. 8.2). So, whereas display rate and tail length are important predictors of male mating success, court quality and, in particular, the central tuft do make a recognisable contribution to it. Whether the neatness of the central tuft provides any evidence of male quality is unclear; however, it does offer the female a structure to hide behind when first she lands and this could contribute to her staying.

The courts of the paradisebird, *Lawes's parotia (Parotia lawesii)*, appear similar in structure to that of Jackson's widowbird, and intermediate in complexity between the simple, unadorned courts of Carola's parotia (*Parotia carolae*) (Gilliard 1969) or of the argus pheasant (Davidson 1981, 1982) and the elaborately constructed and decorated bowers of the true bowerbirds (Ptilonorynchidae) discussed below. Males of Lawes's parotia are found in the forests of New Guinea where they clear individual courts on which they also place pieces of snake skin, chalk, animal fur and the droppings of certain mammals. Pruett-Jones and Pruett-Jones (1988) conducted painstaking observations to determine if these objects were part of the male's display. They also experimentally enhanced some courts with coloured berries and snail shells, typical ornaments used by bowerbirds. They observed that the berries, if edible, were eaten by the resident male and, if not, were thrown off

Figure 8.2
The distribution, mean and S.D. of visits of female Jackson's widowbird (*Euplectes jacksoni*) to display courts in four treatment groups: destroyed, impaired, intact and improved. Only at intact (normal) or improved courts did mean visit durations come close to those for normal approach and copulation. (Adapted from Andersson 1991.)

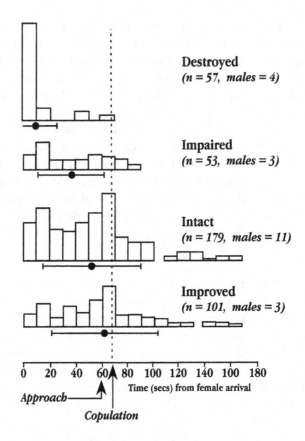

Destroyed
(n = 57, males = 4)

Impaired
(n = 53, males = 3)

Intact
(n = 179, males = 11)

Improved
(n = 101, males = 3)

0   20   40   60   80   100   120   140   100   180
Time (secs) from female arrival
Approach
Copulation

the court by him together with the snail shells. Objects that the males themselves collected were not moved or manipulated as part of the display although males, in the absence of a female, did spend much time rubbing them on the perch above the court which, particularly in the case of chalk, did make the perch more conspicuous. The objects did, however, have a significance for females. In the breeding season they came to courts looking for objects which they took away. Males with more objects received more visits, although this did not apparently result in greater mating success. Pruett-Jones and Pruett-Jones (1988) speculate that the objects are helpful to female reproduction: chalk for egg shell formation and mammal droppings also possibly for some dietary mineral or, more ingeniously, as anti-predator devices to dissuade opossums (*Phalanger* species) or other egg eaters from entering an area apparently occupied by a rival. More obviously, small mammal predators may also be deterred by snake skin (Section 5.2c).

8.5 Bowers and mate assessment

a) The bowerbirds (Ptilonorhynchidae)

The Ptilonorhynchidae is a very small family of only 17 species (Sibley & Monroe 1990), 15 of which build some kind of display structure, the simplest no more remarkable than the court of Jackson's widowbird, the most impressive so elaborate that it was thought by early explorers to have been made by secretive forest people.

The *tooth-billed catbird (*Ailuroedus (Scenopoeetes) dentirostris*) is sometimes called the *stagemaker* because it clears an area of forest floor to create a stage on which it places freshly cut leaves with their pale undersides uppermost (Marshall 1954). Clumping of the courts towards the tops of hills, and competition for sites and leaf ornaments between males (Frith & Frith 1994, 1995) suggests that this is a lek-like display at which females are able to select males on the basis of their displays.

More elaborate is the court prepared by Archbold's bowerbird (*Archboldia papuensis*), which is a carpet of ferns on which are placed snail shells, beetle elytra, lumps of charcoal and of amber-coloured tree-fern resin arranged rather haphazardly (Gilliard 1963, Cooper & Forshaw 1977, Fig. 8.3). On branches overhanging the mat, the male creates dense hanging drapes of, in particular, orchid stems (Gilliard 1963, Frith, Borgia & Frith 1996).

The remaining bowerbirds can be divided into two groups, the *avenue builders* (eight species) and the *maypole builders* (five species). Each group achieves a complexity of architecture and ornamentation that a few years ago was quite baffling. Now, thanks to a revival in observational and several excellent experimental studies, a great deal more is known and understood about both groups.

b) The maypole builders

The characteristic feature of this group is the construction of a column of sticks around the stem of a sapling or small fern to create the central feature of the bower. Its simplest expression is that of Macgregor's bowerbird (*Amblyornis macgregoriae*), which has a maypole two or three times the height of the male composed of a few hundred fine, interlocked sticks in the centre of a circular unadorned moss platform (Fig. 8.4a, Diamond 1982a). The bower of the golden-fronted bowerbird (*Amblyornis flavifrons*) has the same essential elements, but here the arena is decorated with separate little piles of yellow, green and blue fruit (Diamond 1982b).

Figure 8.3
The display area of
Archbold's bowerbird
(*Archboldia papuensis*) is
a mat of ferns on which
are placed small piles of
decorations, including
snail shells and beetle
elytra. Plant material is
also draped over branches
above the court. (Adapted
from Cooper & Forshaw
1977.)

The golden bowerbird (*Prionodura newtoniana*) provides an impressive variant on the maypole design, with twin columns of sticks bridged by a display perch, usually a trailing vine stem. With re-use, these may reach massive proportions relative to the size of the male occupant (Fig. 8.4b), while below the display perch and on the adjoining parts of the stick towers, decorations of lichen, berries and flowers may be arranged (Gilliard 1969, Cooper & Forshaw 1977). The bowers of streaked bowerbird (*Amblyornis subalaris*) (Fig. 8.4c) and the vogelkop bowerbird (*Amblyornis inornatus*) conform to the architectural principle of a single central maypole, but there may be a major additional feature, a hut of sticks enveloping the maypole and opening onto an arena ornamented with piles of decorative objects (Marshall 1954, Gilliard 1969). In *Amblyornis subalaris*, the sides of the hut entrance extend out as two arms to embrace a small court-yard, at the back of which is a broad maypole supporting the hut roof. The maypole is cloaked in a velvety covering of moss and jewelled with embedded flowers and berries which spread out across the courtyard to its front rim (Cooper & Forshaw 1977).

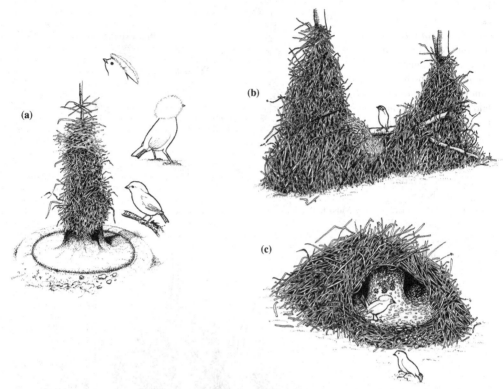

Diamond (1986a, 1987) discovered that in the vogelkop bower-bird there were regional architectural variants. Bowers in the Kumawa Mountains area lacked the hut, consisting only of a single central maypole 2–2.6 m high in the centre of a mat of dead brown moss. Nevertheless, the bower was elaborately constructed and decorated. The hundreds of twigs comprising the maypole were stuck together with a whitish glue of unknown origin and the mat was painted glossy black, apparently with the male's own faeces which, unique to this area, are that colour. Leaning diagonally against the base of the maypole were often carefully placed pandanus leaves and just beyond the perimeter of the mat were objects sorted and piled according to type; typically, several hundred snail shells and brown acorns, with smaller numbers of black-painted sticks, brown stones and a few blackened beetle elytra (Fig. 8.5). These decorations had an estimated weight of 3 kg (Diamond 1987), an impressive accumulation for a bird of only 125 g (Dunning 1993).

The second *Amblyornis inornatus* study site was in the Wanda-men Mountains, some 200 km distant. Bowers here were markedly different in a number of respects. Firstly, the maypole was covered

Figure 8.4
(a) Macgregor's bowerbird (*Amblyornis macgregoriae*) builds a maypole of fine sticks in the centre of a circular moss platform. (Adapted from Diamond 1982a.) (b) The golden bowerbird (*Prionodura newtoniana*) displays on a perch between two large columns of sticks. (Adapted from Cooper & Forshaw 1977.) (c) The bower of the streaked bowerbird (*Amblyornis subalaris*) is a hut of sticks with the maypole supporting it. Piles of decorations adorn the forecourt. (Adapted from Cooper & Forshaw 1977.)

Figure 8.5
(a) Vogelkop bowerbirds (*Amblyornis inornatus*) in the Kumawa Mountains area of New Guinea were found to have no hut-like canopy. Ornaments are laid in piles around the moss mat. (Adapted from Diamond 1987.)
(b) Bowers built in the Wandamen area of New Guinea by Vogelkop bowerbirds have a canopy, with piles of ornaments laid outside the entrance. (Adapted from Hansell 1984.)

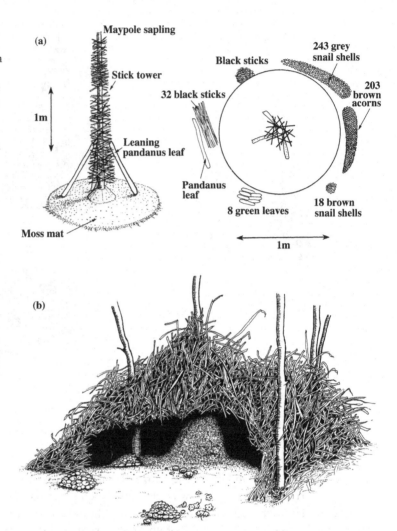

by a hut of sticks which enveloped the moss mat except for a small forecourt; secondly, the moss was not dead and brown, but green and live and without a covering of black paint. The glue holding the maypole sticks together, characteristic of the Kumawa Mountains' population, was also absent. Decorations were plentiful, but again different. In order of commonness, they were black bracket fungi, dark brown or black beetle elytra, then blue, orange and red fruits, and red leaves (Diamond 1987). Diamond noticed that different-coloured objects were generally placed in different parts of the bower, and so he offered Wandamen males coloured poker chips as decorations. In a sample of nine males he found individual preferences, but the overall rankings were blue, orange and purple > red > yellow > lavender > white. The test birds also tended to group their chips with natural objects of the same colour. Colours judged by the test to be more preferred were generally placed inside the hut, with less preferred ones outside (for example, blue 96% inside but yellow only 48% inside). Choices seemed also to have been carefully made; one male initially brought red and lavender chips inside the hut, but later placed red ones with a red fruit outside the front door and removed the lavender chips from the bower altogether (Diamond 1987, 1988).

If bowers are to form the basis of female choice, then females need to be able to compare bowers and the bowers need to reflect some aspect of male quality. Mature *Amblyornis inornatus* males build bowers on ridges, on average only 170 m apart in one study area (Pruett-Jones & Pruett-Jones 1982). Neighbours also appear to be in competition with one another, because they make frequent attempts to damage each other's bowers. However, whether more competitive males secure more matings is not known.

c) The avenue builders

In the three *Sericulus* species, males apparently spend much of the year with no bower and in the fire-maned bowerbird (*S. bakeri*) there is even doubt as to whether a bower is built at all. The avenue of the regent bowerbird (*S. chrysocephalus*) is a relatively simple stick avenue with a few decorations. A mean of 10.3 ornaments was recorded by Lenz (1994), of green leaves, brown fruit and snail shells and, in some, pieces of blue plastic. He found that males stole ornaments and damaged each others bowers, but, although the duration of female visits to bowers was correlated with overall bower quality, it was not correlated with the number of ornaments. In this species the importance of the bower in female choice may be less than in other bower builders; Lenz (1994) found evidence that a male will initially seek out and court a female in the canopy before leading her to the bower.

In the four *Chlamydera* species, there is either no plumage distinction at all between males and females (two species) or a small, pinkish crest on the back of the male's neck which distinguishes him from the female. The avenue is built firmly on a stick platform, generally oriented East–West and painted on the inside with saliva mixed with some macerated vegetation fragments. The platform at one or other ends of the avenue is decorated with numerous objects, typically with large numbers of pale stones, but in the fawn-breasted bowerbird (*C. cerviniventirs*), with small bunches of green berries (Peckover 1970, Fig. 8.6a) or bones (Warham 1957). In the yellow-

Figure 8.6
Avenue bowers:
(a) fawn-breasted
bowerbird (*Chlamydera
cerviniventris*);
(b) yellow-breasted
bowerbird (*Chlamydera
lauterbachi*) – an
East–West avenue with
large twig tufts at each
end; (c) satin bowerbird
(*Ptilonorhynchus
violaceus*), with avenue
oriented North–South.
(Adapted from Cooper &
Forshaw 1977.)

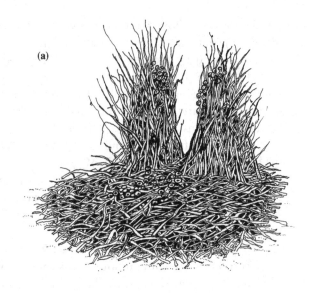

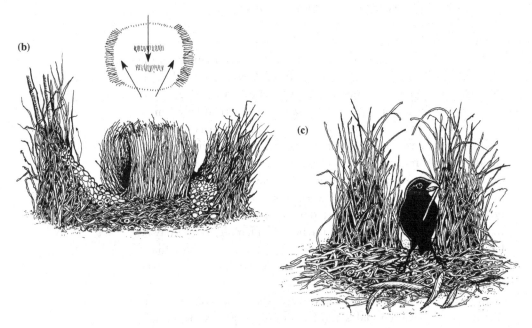

breasted bowerbird (*C. lauterbachi*) there is an additional dramatic elaboration, the presence of two substantial outward-leaning clumps of sticks decorated with stones, shielding the ends of an inner avenue of sticks lined with grass oriented East–West. One such completed bower was composed of about 3000 twigs and 1000 grass stems and was decorated with nearly 1000 stones weighing 5 kg (Gilliard 1969, Fig. 8.6b; male weight 133 g: Dunning 1993).

The spotted bowerbird (C. *maculata*) builds an avenue oriented East–West, but of grass straws not of twigs, giving the walls added transparency (Borgia 1995a). It is decorated with pieces of glass or small stones placed near or inside the entrance to the avenue and sun-bleached mammal bones (typically sheep vertebrae) scattered up to a distance of 2 m away (Fig. 8.7). Borgia and Mueller (1992) and Borgia (1995b) studied the success of male spotted bowerbirds in relation to bower quality. Although bowers in this species are dispersed, males in the study area varied greatly in their mating success, with three out of the 13 males obtaining nearly 60% of the copulations. The success of the males was positively correlated with the number of bower decorations and overall bower quality (Fig. 8.8). Borgia (1995b) suggests that the white bones act first to attract females into the shady area, where the glass ornaments and pebbles then stimulate females when in the bower.

The avenue of the great bowerbird (C. *nuchalis*) is more like that of the satin bowerbird (*Ptilonorhynchus violaceus*) (see Fig. 8.6c), i.e. a narrow avenue, constructed of twigs. The bower of the male satin bowerbird has a well-developed avenue design painted on the inside with the pulp of certain fruit or with ash from burned wood mixed with saliva. The avenue is generally aligned North–South, with the display platform at the North end, the sunny aspect in the southern hemisphere. The platform is covered with yellow straw and leaves and is typically decorated with blue feathers, blue and yellow flowers, cicada skeletons and snail shells (Borgia 1986). Suitable human artefacts may also be included, in one instance, 75 blue plastic objects, including ball-point pen tops and a toothbrush (Vellenga 1970).

P. violaceus bowers generally occur sufficiently close together that a female could, if she wished, walk from one to another to compare them. However, Borgia (1985a), using individually colour-banded birds and video cameras mounted at all bowers, recorded that many visits were actually by other males, which if the resident was away, took the opportunity to damage the bower and steal ornaments. There was a tendency for raiders to be near-neighbours of the victim, but some males turned out to be conspicuous vandals, whereas others were notable victims. Among 30 males followed for one season, one was responsible for 25 destructive raids, compared with eight attacks on his own bower, whereas another male made only six successful raids but had its own bower mutilated 31 times. Importantly, there was a strong correlation between the number of times that a male was raider and the level of aggression it showed towards other males at a feeding station.

Figure 8.7
The avenue of the spotted
bowerbird (*Chlamydera
maculata*) is made of
grass, oriented East–West,
through which the female
is able to observe the male
displaying outside. It is
decorated with stones or
glass and, beyond these,
with mammal bones.
(Adapted from a
well-observed illustration
by Gould 1841.)

There was no correlation between male age and tendency to destroy, but older birds did tend to travel further to conduct raids than did younger males, and to raid individuals at a greater *rank distance* from themselves.

None of this shows, however, that more dominant males have bowers which females regard favourably when choosing a mate. The number of times a male destroyed another bower was not actually found to be significantly correlated with the quality of its own bower, measured as (i) the symmetry of its walls, (ii) size or (iii) density of the sticks forming the avenue, or (iv) measure of overall quality of construction. However, there was a positive correlation between the frequency with which a bower owner destroyed other bowers and the presence of five kinds of decoration in its own (yellow straw, blue feathers, yellow blossoms, cicadas, and human artefacts) (Borgia 1985a). Blue parrot feathers were a particularly favoured target, with males generally taking two or three of the four to six normally present. Borgia and Gore (1986) followed the movement of individually marked blue parrot feathers between bowers for two seasons and showed that the number of feathers in a bower was significantly correlated with the male's stealing activities. However, the number of times a male steals is also positively correlated with the number of times he is a victim of theft (Fig. 8.9).

Understandably, therefore, there was found to be a strong negative correlation between the amount of destruction a bower receives and its quality, measured in the above four ways. All this supports

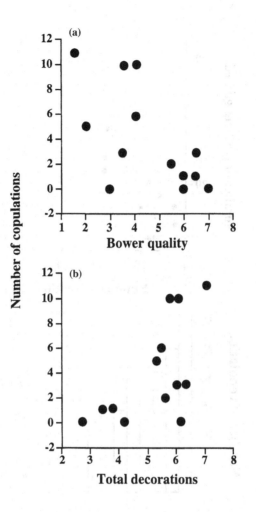

Figure 8.8
Number of successful courtships (copulations) that male spotted bowerbirds (*Chlamydera maculata*) obtained in relation to (a) general bower quality (note: lower values indicate bower of higher quality); (b) total number of decorations. (Adapted from Borgia 1995b.)

the view that there are differences in bower quality that reflect a bird's ability to raid and to protect its own bower and that they are correlated with male dominance in the feeding context. This is given further weight by the study of Hunter and Dwyer (1997) on the satin bowerbird, which confirmed the prediction that, in areas where ornaments are plentiful and therefore of lower value, there is a shift of male behaviour away from stealing ornaments and towards bower destruction.

Borgia (1985b) showed that females did, indeed, visit several bowers and some males obtained many more copulations than others (Fig. 8.10). To determine if mating success was affected by bower quality, he then divided a group of 22 bower owners into two matched groups. In the experimental group, all bower ornaments were removed except a few of the yellow leaves that males

Figure 8.9
Number of feathers gained by male satin bowerbirds (*Ptilonorhynchus violaceus*) in relation to the number lost. (Adapted from Borgia & Gore 1986.)

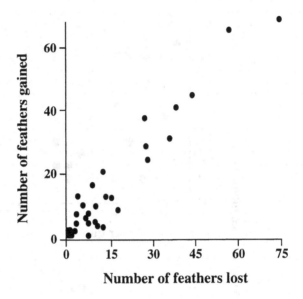

Figure 8.10
Male satin bowerbird (*Ptilonorhynchus violaceus*) mating success is very varied, with the majority of matings obtained by just a few individuals (each column represents an individual male). (Adapted from Borgia 1985b.)

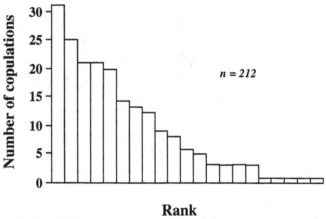

tend to hold when courting the female, while controls had all ornaments left in place. The result was significantly fewer copulations in the experimentally robbed group compared with the decorative controls. Even among the controls themselves there was variation in mating success. Analysis of this in relation to decorations showed that female preference was correlated with the presence of snail shells, blue feathers and, to a lesser degree, yellow leaves (Table 8.2). Mating success was also positively correlated with all four measures of bower quality (wall symmetry, size and density of the sticks, and overall quality of construction), showing that the construction of the avenue was also important to females (Borgia 1985b).

Table 8.2. **Analysis of male satin bowerbird mating success in relation to numbers of nine different ornaments**

Decoration	1980 ($n = 11$)	1981 ($n = 22$)
Yellow leaves	NS	S
Yellow straw	NS	NS
Blue feathers	S	S
Snail shells	S	S
Blue blossoms	NS	NS
Yellow blossoms	NS	NS
Cicada skins	NS	NS
Man-made objects	NS	NS
Natural objects	NS	NS

The analysis shows that only blue feathers and snail shells had a significant effect in two study years. (Adapted from Borgia 1985b.)
S = significant, NS = non-significant.

Females, it seems, are making quite sophisticated assessments of bower quality which are reflected in the males' construction behaviour. Borgia, Kaatz and Condit (1987) used this to examine in more detail the choice of flowers as decorations. They found that only seven different flower types accounted for 93% of all those appearing in bowers. These varied in colour, but differed significantly from what was locally available; purple and dark blue were over-represented, yellow and white under-represented, whereas red was totally absent (Fig. 8.11, Borgia *et al.* 1987). By placing near to bowers, locally occurring flowers and novel flower types created by dyeing white hibiscus flowers either red or blue, they showed that blue flowers were more readily taken into bowers than white ones and that, of the dyed hibiscus flowers, the blues were about as acceptable as the natural white, but reds were ignored.

d) Functional design and bower evolution

All the bower designs (court, mat, maypole and avenue) are apparently display devices serving to enhance a male's vocal and visual behaviour as it finally persuades the female to mate. However, the avenue design also offers an opportunity for concealment, as does the presence of a hut around the maypole in bowers of two of the *Amblyornis* species. These could give *proximate benefit* to a female in the form of protection to the courting pair from side or overhead attack. This predation protection hypothesis is supported by

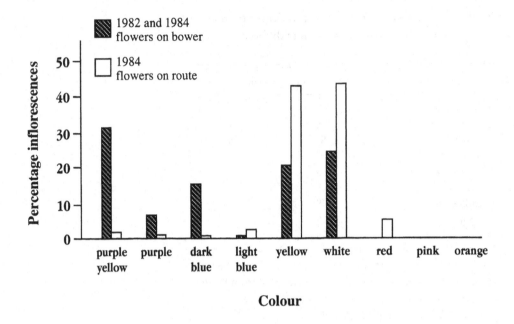

Figure 8.11
Male satin bowerbirds (*Ptilonorhynchus violaceus*) show a preference for certain flower colours. The frequency of flowers on the bowers (dark bars) shows that blue and purple flowers are over-represented and yellow and white flowers under-represented when compared with flower frequency estimated by human transect sampling route (pale bars). (Adapted from Borgia *et al.* 1987.)

evidence of a broad association between open forest habitats and enclosed avenue and hut bowers, and between dense forest habitats and more open bower designs such as the maypole of *Am. flavifrons* and the mat of *Ar. papuensis* (Borgia, Pruett-Jones & Pruett-Jones 1985, Table 8.3).

The question of the evolutionary origin of bowers was first given a coherent explanation in the *transfer* hypothesis of Gilliard (1963, 1969). He proposed that, in the ancestral situation, brightly coloured males prepared simple display courts, as is seen in paradisebirds like *Diphyllodes* (*Cicinnurus*) *magnificus*. Selection then favoured a progressive transfer of the display from the feathers to the construction of the display area. In support of this hypothesis, Gilliard (1963) pointed out the inverse relationship between male crest size and bower complexity in the maypole-building group. *Amblyornis macgregoriae* and *Am. flavifrons* have the simplest bowers and the males have large orange crests; *Am. inornatus*, as its name suggests, has no bright plumage but the most complex bower; *Am. subalaris* is intermediate in both respects. With the display now transferred away from the male, its ornamentation could then proceed relatively uninhibited by any additional risk of attracting predation, because the male could quickly disassociate itself from its conspicuous appearance. However, in proposing the transfer hypothesis, Gilliard (1969) had supposed that paradisebirds and bowerbirds were close relatives and that ancestral bowerbirds

Table 8.3. **The relationship between bower structure and habitat types**

	Open bower	Closed bower
Open forest		*Chlamydera*
		Ptilonorbynchus
		Amblyornis inornatus
Closed forest	*Archboldia*	*Amblyornis subalaris*
	Prionodura	
	Ailuroedus (Scenopoeetes)	
	Amblyornis macgregoriae	
	Amblyornis flavifrons	

Closed bower designs tend to be associated with open habitats, and vice versa. Modified from Borgia *et al.* (1985).

would have had bright plumage. DNA hybridisation studies conducted since do not support this, instead showing proximity to the lyrebirds (Menuridae) (Sibley *et al.* 1988); lyrebirds show several display characters in common with bowerbirds, including the preparation of display courts (Section 8.4), vocal mimicry, and a long delay in males before the onset of maturity (Ligon 1999).

The phylogeny within the bowerbird family has been inferred from base-pair sequences in a region of the mitochondrial cytochrome *b* gene (Fig. 8.12, Kusmierski *et al.* 1997). Fitting the features of bowers and plumage attributes of the two sexes to this cladogram supports the overall view that the maypole builders together with *Archboldia* and *Ailuroedus (Scenopoeetes)* form a monophyletic group, as do the eight species of avenue builders in the three genera *Sericulus*, *Ptilonorhynchus* and *Chlamydera*. The use of sticks may have been a single evolutionary event for all bower builders and have been lost in *Ailuroedus* or possibly have evolved independently in avenue and maypole builders (Kusmierski *et al.* 1993).

The transfer hypothesis (Gilliard 1963) in this phylogeny (Fig. 8.12) is indeed supported by the avenue-building group; however, in the maypole group its predictions are not upheld. The dull plumage of *Ailuroedus* may represent the ancestral condition (Kusmierski *et al.* 1993, 1997), and there appears to be little consistency in the combinations of bower features, crests and bright body plumage. Kusmierski *et al.* (1997) conclude that, in general, display traits of bowerbirds make poor characters for tree reconstruction and that in this family these behavioural traits are labile and show little evidence of phylogenetic inertia.

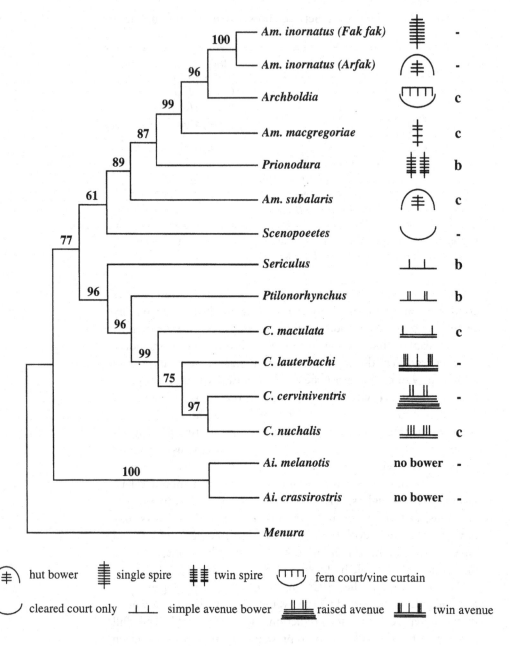

Figure 8.12
Phylogeny of 14 bowerbird species rooted with the superb lyrebird (*Menura novaehollandiae*, Menuridae). Numbers at the nodes indicate the percentage of times each was confirmed through bootstrap replication, hence the confidence attached to it. The type of bower is shown as a hieroglyph for each species, with the complexity/size of the bower represented as an increase in the number of lines. Plumage dimorphism between the sexes is shown as: c = crest, b = bright plumage, – = no dimorphism. (Adapted from Kusmierski *et al.* 1997.)

The position of *Ailuroedus* in the molecular-based phylogeny of Kusmierski *et al.* (1993, 1997) supports the view that dull male plumage and absence of bower building are ancestral traits. What, therefore, did the bower evolve from? In *Ailuroedus crassirostris*, pairing is monogamous and males help to build the nest (Gilliard 1969). This gives support to the *nest hypothesis* for the origin of the bower. Collias and Collias (1984) claim similarities between the construction of the male bower of *Chlamydera maculata* and *C. cerviniventris*, described as, respectively, an avenue of twigs lined with grass stems or with fine rootlets, and the two-layered structure of the female nest. Borgia *et al.* (1985) reject the nest as the ancestor of the bower, because the arrangement and predominant orientation of the twigs are different and because nests are built not on the ground but in trees. They argue that, since nests are built by females before visiting the bower, the males' reproductive behaviour is clearly not necessary to induce reproduction.

Discussion on the ancestry of the bower has also considered whether the fruits used as ornaments in the bowers are homologous with those used in courtship feeding. In several bird families, males feed females as part of courtship behaviour (Skutch 1976). Diamond (1982a) proposes that, in this ancestral condition, the nest and associated male behaviour are necessary to bring the female into a reproductive state. The display of a male *Am. flavifrons*, in which the male picks up a fruit from the bower and holds it before the female, should therefore be viewed as originating from courtship feeding at the nest. Further selection for fruits as ornaments rather than food could then have led to the inclusion of inedible objects of more varied kinds. Borgia *et al.* (1985) also reject this interpretation, particularly on the ground that the objects which males display, whether held in the beak or not, are almost invariably inedible.

Borgia (1995a) interpreted the thin grassy wall of the avenue of *C. maculata* as an evolutionary modification to allow the female to stand in the avenue and observe the male display behind the protection of a diaphanous curtain of grasses (see Fig. 8.7). He argues that, in this species, although mating success is correlated with bower ornamentation, the relative lack of direct male–male competition resulting from widely spaced bowers (Borgia & Mueller 1992) has led to selection for vigorous male display to females as part of mate assessment. He explains the change in orientation of the bower to an East–West alignment by a *threat reduction* hypothesis, the broadness of the avenue and transparency of its walls providing a female with security while observing the male as he runs at the wall with rapid and vigorous movements from 3–4 m away. According to this explanation, the avenue design provides *proximate benefit*, with females choosing bowers that provide greater protection against male aggression.

Borgia (1995c, 1997) develops this threat reduction hypothesis, noting that whether a bower is avenue or maypole, it still provides a method of separating male from female. In *Am. macgregoriae*, for example, a female will land on the maypole and then move down onto the arena. When she lands, the male initially positions himself on the opposite side of the maypole to her, in a manner reminiscent of a male Jackson's widowbird on its court (Andersson 1991).

Bowerbird species with courts rather than bowers face the same problems of male aggression, but have solved them in different ways. Tooth-billed catbird males show aggression in their courtship, but hide behind trees on their courts while calling females down. In Archbold's bowerbird, males displaying on the court adopt a prostrate attitude which may serve to reassure females of their ability to escape in case of male attack (Frith *et al.* 1996, Borgia 1997).

Despite this support for the threat reduction hypothesis, there remains the problem of explaining the evolution of a device built by males to make female less accessible to them. This is countered by Borgia (1995c, 1997), who argues that males building such structures were compensated by the increased attendance of females, protected against the risk of forced copulations. The security provided by these bower structures does also allow the female a better opportunity to observe the male displays, including the bower ornaments. The ornaments, in turn, can be explained as resulting from sexual selection of some kind operating through female choice.

8.6 Bowers and sexual selection theory

The passive choice hypothesis proposes that females are not, in fact, exhibiting choice at all but mate with more conspicuous males because they are more likely to be located first. This can apparently be rejected in the case of the satin bowerbird, for which the camera records of Borgia (1986) confirm that females visit several bowers before mating.

For the proximate or direct benefit hypothesis to be supported, the female must obtain direct fitness advantage from the display itself. The quality of male plumage might give evidence of a male's freedom from parasites or testis development, but there is no evidence that the bower itself provides such information. However, the proximate benefit hypothesis is supported by weak evidence that bowers might provide some protection for females from predators (see Table 8.3) and by stronger evidence of its role in threat reduction (Borgia 1997) by providing protection from male attack.

The elaborate ornamentation of bowers shown by both avenue and maypole builders accords with both Fisher's runaway and good genes predictions. In the former, the selection of the trait should be arbitrary and run away till checked by natural selection, for example from predator pressure. The differences between the satin and the spotted bowerbirds in the avenue width, composition and orientation, and the spatial arrangement of ornaments by spotted bowerbirds into bones for more distant and glass or stones for close-range attraction (Borgia, 1995a, 1995b) all argue against the arbitrary nature of bower design features. For the runaway hypothesis to be supported, bowers should have a high cost (Lande 1981), but this is also true of the *handicap* interpretation of the good genes hypothesis. Diamond (1986b) argues that bower building costs were high; however, Borgia (1993) disagrees, so the case for the handicap and runaway hypotheses remains inconclusive, while a good genes explanation cannot be dismissed.

A specific form of the good genes interpretation of the role of the bower is the *marker hypothesis* (Borgia *et al.* 1985). This proposes that, in a lek-like concentration of bowers where neighbours steal ornaments from and damage the structures of neighbours, the quality of the bower serves as a marker of male quality which females can readily compare. Ornament stealing and the damaging of neighbours' bowers are not only, as already noted, characteristic of male satin bowerbirds, but are also reported for the regent bowerbird (Lenz 1994) and the maypole-building *Am. inornatus* (Pruett-Jones & Pruett-Jones 1982).

Borgia (1986) and Borgia *et al.* (1987) argue that, in the satin bowerbird, the specific male preferences for rare blue flowers and their intense competition for scarce blue parrot feathers favour the good genes explanation above the others. The quality of their bowers is also correlated with male dominance in a feeding context (Borgia 1985a, 1985b). This evidence all supports the view that a female will be able, on the basis of the bower, to select a competitive male; what remains to be demonstrated is that his offspring are better survivors or his sons more successful than those of males with poorer bowers.

This good genes interpretation is less easy to apply to the spotted bowerbirds, the males of which are widely spaced and little prone to ornament stealing or bower damage from rivals. The preference for more ornamented bowers could be explained by the runaway hypothesis; however, Borgia (1995b) points out that the long occupation by males of the same site could allow females to recognise males as good survivors, or the enhanced intensity of male courtship in the spotted bowerbird compared to that of the satin bowerbird could indicate that females are selecting for male vigour, both good genes interpretations for displays in this species. Humphries and Ruxton (1999) further argue that, in species with more widely spaced bowers, where travel time between bowers will be longer, females may select for less complex male displays, giving them the proximate benefit of reduced costs in terms of search time and demands upon their short-term memory, in terms of display details.

The quality of the bower of the satin bowerbird improves with age (Borgia 1985b, 1986). This could, therefore, provide evidence of a male's ability to survive The delay in the onset of reproductive age of bowerbird species means that all displaying males have already shown some capacity to survive, raising the question of why they should not cheat by maturing early. This was tested by Collis and Borgia (1993) using testosterone implants to induce the early expression of adult plumage in juvenile satin bowerbirds. Experimental males were found to be no better in bower construction or other display characters than untreated controls and are tolerated less by adult males; they seem to be disadvantaged by their accelerated maturity in not having time to learn the displays critical for later reproductive success. The indication that bowers could be an honest signal of male survival through the expression of displays requiring long experience (Borgia 1993) is interesting in suggesting a mechanism by which ornament might become elaborated.

8.7 Beautiful bowers?

The Ptilonorhynchidae, not uniquely, but overwhelmingly, are the builders of bower-like structures. To explain this, Diamond (1988) suggests that the lack of mammalian predators and the scarcity of competing mammalian frugivores have allowed sexual selection for elaborate, fixed ground displays to evolve relatively unchecked. The occurrence of an independently evolved display with court objects in the paradisebird Lawes's parotia (Pruett-Jones & Pruett-Jones 1988), also found in New Guinea, lends support to this hypothesis. However, as we have seen, a bower-like display is created by an African savannah species, Jackson's widowbird (a relative of the weavers) (Andersson 1989) and indeed by a lake-dwelling cichlid fish, the females of which choose males on the basis of the size of the sand pile they heap up (McKaye, Louda & Stauffer 1990).

All these may, at least in part, be the consequence of female choice for good genes, but what good genes are these? The argument developed by Borgia (1995b) is in terms of selection for male vigour. An example of such a hypothesis was formulated by Hamilton and Zuk (1982). It states that feather and skin ornaments displayed by male (birds) to females would be an honest signal of body condition and so an indication of freedom from parasitism, a character that could be heritable. Similarly, the *fluctuating asymmetry* hypothesis proposes that high-quality individuals will produce large, symmetrical ornaments, poor-quality individuals the reverse (Evans & Hatchwell 1993). Møller (1992, 1993) has shown that female barn swallows (*Hirundo rustica*) prefer males with long, symmetrical tail pennants and that these do reflect body condition. So, can bowers be explained simply as a signal to females that males are in healthy, vigorous condition? The stone-carrying behaviour of the black wheatear (*Oenanthe leucura*), although not an example of bower building, illustrates this principle; females invest more in reproduction when mated to a male that is a relatively impressive stone collector (Moreno *et al.* 1994).

For satin bowerbirds, the evidence is clearly that bower quality is an indication to females of male vigour. Not only number of ornaments but also symmetry of the bower can be interpreted as measures of a male's ability to acquire ornaments and protect the bower from damage. Nevertheless, the degree of complexity in some avenue and maypole bowers makes this explanation seem unsatisfactory. There is an inherent contradiction in demonstrating in bowerbirds the need for several years to acquire display skills (Borgia 1993) and showing that bower designs continue to change with age (Borgia 1995a) yet females judge quite gross features of the bower such as number of blue ornaments.

Diamond (1982b, 1988) argues for an aesthetic sense in bowerbirds, while Borgia *et al.* (1987) take a more conservative view, arguing that differences in preferred ornaments between species speak of a lack of general aesthetic criteria and that such a hypothesis is anyway, untestable. Darwin (1871) was unequivocal: '... the playing passages of bower-birds are tastefully ornamented with gaily-coloured objects; and this shows that they must receive some kind of pleasure from the sight of such things'. For most of the twentieth century it was unacceptable for scientists to express such views, largely because they did seem inaccessible to scientific investigation, but Griffin (1981) reopened the scientific debate on animal consciousness. Whiten and Byrne (1988), testing the hypothesis of social manipulation by primates, have further shown that a careful compilation of anecdotal observations can constitute a powerful argument for self-aware behaviour.

Cronin (1991) contrasts the view expressed on animal ornamentation by Darwin in *The Descent of Man* with that of his contemporary and co-founder of the theory of evolution by natural selection, Alfred Russell Wallace, i.e. that, because male adornments like the peacock's tail are useless, females must prefer them for their beauty (Darwin 1871), with the argument that male adornment is an expression of his vigour and condition and therefore female choice is functional and utilitarian (Wallace 1889). Cronin characterises this argument as one of good taste versus good sense. Her resolution is that, whereas Wallace's position is currently supported by the 'good genes' argument, that of Darwin is justified by Fisher's runaway theory. However, this seems to avoid the debate about whether males and females have taste in the sense that Darwin declares for the bowerbirds, i.e. that they *receive some kind of pleasure* from seeing the bower. This challenge is, however, directly addressed by O'Hear (1997), whose general thesis is that certain human attributes, appreciation of beauty included, are beyond explanation in evolutionary terms and that no animal, excluding only ourselves, expresses that sensibility. However, in doing this he fails to consider bower construction, where the ornament is expressed through the animal's behaviour, concentrating on bird plumage examples such as the peacock's tail. Thus, he dismisses rather readily any animal's ability to appreciate beauty by citing criteria which it seems possible that bowerbirds may satisfy. Nevertheless, he does tend to emphasise that animal responses to adornment in others 'look much more like genetically programmed responses...', that animals cannot judge something to be beautiful in a context detached from that of its function, and that animals over time cannot come to a view that certain properties in a display are beautiful which they failed to understand before. Evidence currently available through bowerbird studies suggests that O'Hear may be wrong in all these particulars. That is the evidence of male bowerbird *apprenticeships*.

It takes about seven years in satin bowerbirds for a male to attain adult plumage, and bower-building skills apparently require an extended period of learning; several years are also apparently needed to learn stage making in *Ailuroedus dentirostris* and maypole construction in *Prionodura* and *Am. macgregoriae* (Diamond

1986b). Vellenga (1970) records immature satin bowerbird males observing adults, entering their bowers in the owner's absence, and sometimes applying saliva without pigment to the walls. Birds no more than two years old were seen to work together to make a basic platform, while other young males repeatedly constructed and dismantled recognisable but incomplete avenues. Borgia (1995b) records male/male displays in the spotted bowerbird and suggests that males may learn display elements from one another. Borgia (1995a) records that the walls of the avenue of the satin bowerbird show a more pronounced tendency to bend in at the top when built by older birds; so there are age-related changes, which could be regarded as maturing taste as well as skill.

Diamond (1987) records marked regional differences in bower architecture in *Am. inornatus* and speculates that these differences are culturally determined. As evidence, he cites, firstly, the presence in two areas with clearly different architecture of many of the same materials for construction and decoration; secondly, the period of four to seven years during which males practise bower making before they eventually breed; and, thirdly, the occurrence of decorative idiosyncrasies shared between near neighbours, such as the use of beetle heads or butterfly wings. Regional differences are now also reported in court preparation by *tooth-billed catbirds (Frith & Frith 1994) and may exist in bowers of Archbold's bowerbird (Frith *et al.* 1996). So, when male bowerbirds show, as they do, apparent fastidiousness in their arrangement and rearrangement of ornaments (Diamond 1988), this could be viewed as evidence of the 'disinterested' and 'reflective' judgement that O'Hear (1997) demands, possibly also of a consideration of the effect the display might have on females, an important element of consciousness. None of this is enough, but the point is that very much more careful developmental studies coupled with experimental manipulation seem worth while on these species.

If the case were made, then the challenge would be to explain how the ability to judge beauty might have evolved through natural selection. A satin bowerbird display may, through the number of ornaments and symmetry of the avenue, provide females with a pocket calculator assessment of male competitiveness. However, if it really takes more than five years to acquire adequate courtship performance, then it certainly seems possible that the bower contains a lot more information than that. This could include information on manipulative skill, or even on mental abilities such as learning capacity and the ability to judge spatial relationships. In the case of regional differences in court construction by

Ailuroedus dentirostris (Frith & Frith 1994) or bower construction by *Am. inornatus* (Diamond 1988) it might indicate differences of 'culture' and arguably, in the case of individual idiosyncrasies in ornament choice by *Am. inornatus*, of 'inventiveness' (Diamond, 1987, 1988). One or more of these attributes might benefit a female that chose them.

CHAPTER 9

The evolution of nest building

9.1 Introduction

In Chapter 2 it was argued that nest building has been a feature of the biology of birds since their origin, and has had a key role in shaping the relationship in bird reproduction between parents and offspring. The prevalence in birds of biparental care (Ligon 1993) and the universal occurrence of egg laying (Blackburn & Evans 1986) are apparently important manifestations of this. During the evolution of bird nests, designs have become diverse, particularly in the passerines, the order that contains 59% of all living species of birds compared with 41% for all the 22 non-passerine orders put together (Sibley & Monroe 1990). As there is no fossil record of bird nests, can the pattern of diversification of bird nests be mapped? Has the appearance of new nest designs altered the rate of evolution (in particular of speciation)? Do nest characteristics such as those described in Chapter 5 provide reliable evidence for the reconstruction of bird phylogeny? These are matters for consideration in this chapter.

Collias (1997) distinguishes three types of nest based on their overall architecture: *hole*, *open* and *domed* nests. Examining the handful of families considered to be among the most primitive of the passerines, i.e. (in the sub-order Tyranni) the Eurylaimidae (broadbills), Formicariidae (ground antbirds), and Rhinocryptidae (tapaculos), and (in the sub-order Passeri) Menuridae (Australian lyrebirds) (Sibley *et al.* 1988, Sibley & Monroe 1990), he finds all three basic nest types represented. The Eurylaimidae universally build domed nests, the Rhinocryptidae predominantly burrow nests, and the Formicariidae open cups. Collias (1997) concludes that the diversity of bird nest design evolved early, so early, indeed, that the ancestral form in the nest of the order Passeriformes cannot be deduced.

The explanation of Collias (1997) for this early diversification of nests is based partly on the nature of the birds themselves and partly on the ecological opportunities offered to them. Mobility gave them access to the great diversity of new habitats that were appearing in the late Tertiary due to climate change. Small body size in passerines and an ability to place nests in a wide variety of locations gave rise to diversification of nests. This, suggests Collias, may even have contributed to the rapid speciation in birds. Each of these claims – the importance to nest evolution of small body size, access to new nest sites, new habitats in the late Tertiary, and the role of nests in speciation – deserves examination.

The time scale of avian radiation is a matter of current debate. One view is that mass extinctions occurring as a result of the Cretaceous–Tertiary (K–T) boundary event 65 million years ago produced an evolutionary bottleneck in birds and mammals, so that the diversity we now see is the result of rapid Tertiary radiation (Feduccia 1995). Consequently, the Passeriformes, which now comprise nearly 60% of living species of birds, appeared since the late Oligocene, less than 40 million years ago. This hypothesis, which is based largely on a fragmentary fossil record, is the one supported by Collias (1997). However, it is a view contradicted by three recent studies, which use molecular evidence of differences in living species to estimate divergence times for different lineages: those of Hedges *et al.* (1996) and Cooper and Penny (1997), and by Padian and Chiappe (1998) in a review covering both molecular and fossil evidence.

Hedges *et al.* (1996), using a suite of nuclear and mitochondrial genes, calculate a date for the diversification of avian orders of about 100 million years ago, in the mid-Cretaceous. This diversification they associate with coincident fragmentation of emerging land masses, a trend likely to promote speciation. Cooper and Penny (1997), using similar techniques, calculate a divergence point slightly earlier and identify 22 avian orders – including the Passeriformes – as surviving the K–T boundary event. Padian and Chiappe (1998) suggest an origin of the class Aves earlier than 150 million BP. None of these studies supports the idea of an evolutionary bottleneck at the K–T boundary and they all portray a slower rate of radiation in birds than envisaged by Feduccia (1995).

If a diverse range of bird orders did survive the K–T event, then they had certain adaptive attributes that were notably lacking in the dinosaurs. One of these might have been the building of protective nests in which parents incubated the eggs; however, this seems unlikely to have been a primary cause. The cometary or asteroid impact hypothesis (Alvarez *et al.* 1980) envisages a single cataclysmic event centred possibly off the Yucatan peninsular of Mexico. The impact would have flung massive quantities of dust into the atmosphere, blocking solar radiation, largely stopping photosynthesis for six months and precipitating a global winter lasting two or more years, before the atmosphere could clear and vegetation regenerate. Some animal builders would certainly have been highly adapted to survive such severe but transitory conditions; burrow-dwelling rodents and ants that stockpile seeds in granaries are obvious examples. In these circumstances the ability of birds to build nests for breeding during this critical period would probably have been of little significance compared with an ability of small birds simply to roost in cavities which conserved heat, or to fly to locations where conditions were more favourable. Cavity nesters are present in a wide range of bird families (Collias & Collias 1984) and, in the non-passerine orders, Coraciiformes (which includes all kingfishers), Piciformes (woodpeckers and allies), and Trogoniformes, cavity roosting and nesting are virtually universal (Ligon 1993), perhaps giving such species a better chance of survival. At least some dinosaurs were building protective nests (Section 2.2), but many were too large to find suitable shelters and lacked the mobility of birds to evade the effects of a widespread catastrophe.

The diversity of nest designs in smaller birds compared to larger ones claimed by Collias (1997) is in general supported in Chapter 5. Small nests in particular may incorporate several protective principles: concealment, mechanical protection, inaccessibility and insulation; in larger bird species, these functions can, to a greater degree, be underaken by the eggs and chicks themselves. Evidence in Section 5.2a supports the view that nest concealment devices are largely a characteristic of smaller nests. This pattern of less complex and diverse building in larger species is quite evident in mammals, which show a greater range of body sizes than birds. Nest building and burrowing are commonplace in smaller species of mammals and virtually absent in large species. Small nest size in birds probably offers a greater range of sites than larger nests; small nests may be secreted in crevices or placed on the fine extremities of vegetation, sites less accessible to larger species. The largest nests are also mechanically unsuited to being supported only from above. A variety of factors, therefore, may have given rise to greater nest diversity in small compared to large birds.

More interesting, however, is the claim of Collias (1997) that nest diversity, for whatever reason it arose, can 'help explain' the species diversity of birds. How could nest building promote speciation? Allopatric speciation occurs in situations where geographically separated parts of an original population diverge phenotypically through selection or drift to a sufficient extent that, on removal of the barrier, interbreeding is no longer possible or hybrids are selected against. A special case of this is speciation by the *founder effect*; that is, where a small peripheral population becomes isolated in a novel environment. This may lead to disruption of co-adapted gene complexes in the parental population, resulting in sometimes rapid evolution towards a new adaptive peak (Carson 1982).

An extreme example of rapid speciation is the appearance of about 500 new species of cichlid fish in Lake Victoria in less than 12.5 thousand years (Seehausen, van Alphen & Witte 1997, Gails & Metz 1998). Males of these cichlids are characterised by bright and varied colour morphs; the behavioural effect of female choice for different male morphs within a species appears sufficient to establish reproductive isolation, even in sympatric populations. Competition for food between the incipient new species seems to select for feeding specialisations, which are able to evolve rapidly because of the large number of independent elements in the jaw apparatus of these fish. Behavioural flexibility and learning ability aid this process of specialisation.

The role of behaviour in exercising a pioneering influence on the evolution of organisms was recognised by Bateson (1988). He pointed out that behaviour ensures that animals are not passive recipients of selection pressures but, by making choices, affect the selection pressures to which they are exposed. In addition, through their high degree of mobility, animals can place themselves in novel conditions which may reveal new kinds of heritable variation, which could contribute to speciation. So, in the context of nest building, an innovation in design or materials might arise in an isolated population, which can be exploited by the choice of a new type of nest site. Also, choice of a new kind of nest site could permit a sudden shift in habitat range, cutting off gene flow between the pioneers and the rest of the population. Fixation of the new nest design could then accelerate evolutionary change by allowing further innovations in the new habitat previously selected against because they were incompatible with the original architectural specifications. If at this point sympatry were restored, speciation might occur because matings between individuals with distinctive nest-building genotypes produced offspring with inconsistent building behaviour (Hansell 1996a).

The adoption of a new nest site might occur rapidly through the learned process of behaviour imprinting. By this mechanism the novel nest site choice of a parent could become the preferred nest site choice of the offspring; this could isolate a subpopulation, leading to allopatric or even sympatric speciation. This effect is suggested by an anecdotal observation on female paper wasps (*Polistes annularis*). Females forced in an experiment to construct nests in an artificial site raised daughters which the following year tended to choose the artificial site to build their nests. The impression of the importance of learned nest site preference in this species is reinforced by the observation of distinctive local nest site preferences (Wenzel 1996).

The flexibility that behaviour offers an animal in its interaction with new or changing environments therefore could produce rapid rates of evolution. Since more complex behaviour is associated with greater neural complexity, if behaviour were an important spur to evolutionary change, the rate of evolution should be positively correlated with relative brain size. This is, indeed, demonstrable for terrestrial vertebrates (Bateson 1988), supporting the view that evolution of new nest types in birds could be rapid and, consequently, that nest building may have had a significant role in their speciation.

The greater species diversity in the order Passeriformes compared with the other 22 orders of birds put together was examined by Riakow (1986) to assess whether there was any *key adaptation* in the passerines to account for the species richness of the order. Riakow examined various candidates, including perching mechanisms, feeding anatomy, and the varied syrinx apparatus, but concluded that no key character is evident; indeed, that the question itself may in part be a taxonomic artefact. Interestingly, Riakow (1986) does not even consider the nest as a possible key adaptation. Although this may have been an oversight, I find it hard to make the case for him. In the social insects such as wasps, nests appear to have had a significant role in speciation (Hansell 1996a), but, as has been stressed already (Section 4.1), very little of an adult bird's interaction with its environment involves building a nest compared with, say, foraging. The marked differences in beak morphology of Hawaiian honeycreepers (Drepanidini) compared to their nest differences provide evidence for this (see Fig. 4.1). However, as seen in the African lakes cichlid fish example, sexual selection can be a powerful force for speciation because of its direct effect on gene flow None of this excludes nest building from a role in bird speciation, but, on present evidence, its diversity in passerines could equally well be explained as the consequence of speciation or as the cause of it.

9.2 Innovations of design and of technology

Innovations in nest building can broadly be divided into two types, innovations of design and of technology. In the former, the pattern of the building behaviour is altered to produce new architectural features; in the latter, the key behavioural change is in the selection of materials. The effects of technological change could be the more profound because the new material might open up a whole range of new design options, leading to a radiation in the architecture. This is well illustrated in the wasps (Vespidae), for which paper supplanted mud as a nest building material probably in the first instance because it allowed the nest to be attached by a fine stalk to a new range of sites. This new material proved so strong and versatile that it led to the evolution of a new range of comb and envelope designs, and a considerable increase in nest size and hence colony size (Hansell 1987a). With an increase in nest size came new arrangements of comb and envelope (i.e. innovations in design), accompanied by some innovations in composition of the paper (i.e. innovations of technology) (Wenzel 1991). In the Polistinae (Vespidae) this has resulted in a pattern of evolution in which a phylogeny based on nest characters shows reasonable concordance with that based on the wasps themselves, suggesting that the nest in these organisms has, indeed, had some role in speciation (Hansell 1996a).

An association of some kind between nest building behaviour and speciation in birds is demonstrated by the study of Winkler and Sheldon (1993) on the swallows and martins (Hirundinidae). This provides an example in which a single technological change has resulted in new architectural diversity. Analysis on 17 species of 15 (i.e. essentially 100%) of genera generated three fundamental clades: *Hirundo* and allies, core martins (*Progne*, *Riparia*) and the African sawwings (*Psalidoprocne*). The sawwings are burrow nesters, as are *Riparia* species, whereas *Progne* use available cavities. The technological revolution was a change from mud as spoil from excavation to mud as a construction material. This innovation appears to have occurred once only in this lineage and from it stemmed the diversity in constructed nests seen in *Hirundo* clade (Fig. 9.1) (Winkler & Sheldon 1993). Construction with mud translated nest sites away from soft sand or soil exposures to cliff ledges or similar sites where the adhesive properties of the mud allowed the design innovation of attached nests. Initial designs were a cup or half cup partly supported from below (barn swallow, *Hirundo rustica*, and rock martin, *Ptynoprogne fuligula*); from this evolved

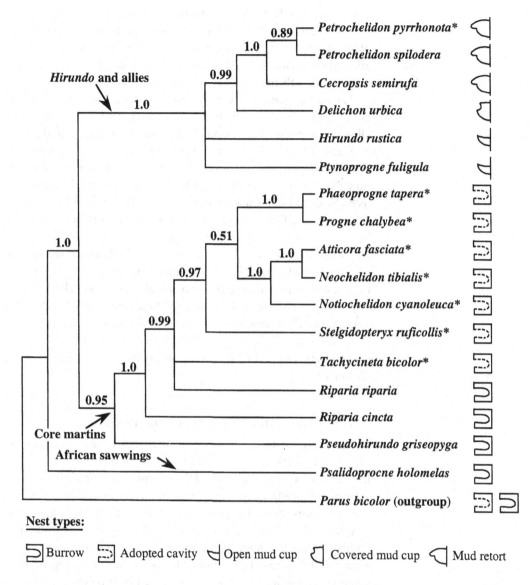

Nest types:

⊃ Burrow   ⊂∷⊃ Adopted cavity   ⊲ Open mud cup   ⊲ Covered mud cup   ⊲ Mud retort

Figure 9.1

Majority rule consensus tree for the phylogeny of swallows (Hirundinidae)
based on DNA–DNA hybridisation. Constructed mud nests appear to have
evolved once: *Hirundo* and allies. Values at the nodes indicate (with a
maximum value if 1.0) the confidence that may be attached to them. (Adapted
from Winkler & Sheldon 1993.)

nests supported from the nest rim to a rock overhang (house martin, *Delichon urbica*); the final design elaboration being the appearance of an entrance tube (*Petrochelidon* spp.). This pattern does suggest an active role for technological innovation in the speciation process, possibly through the colonisation of new breeding habitats.

The building material innovation of sticks to create the maypole architecture shown by the bowerbird genus *Amblyornis* may have allowed the design innovation of stick avenues to evolve in the satin bowerbird *Ptilonorhynchus* and related genera (see Fig. 8.12), although independent evolution of twig use is a possible interpretation (Kusmierski *et al.* 1993). Of much greater significance as a technological breakthrough has been the incorporation of spider silk into nests. The importance of this to small birds has been noted by a number of authors (Snow 1976, Collias & Collias 1984, Hansell 1984). Designed as a material with high tensile strength, it has functioned in the role of attachment for and strengthening of bird nests where tensile stresses are a major problem. This technology is one that has been adopted many times independently (Section 5.4e), probably as much for convenience during construction as for its tensile strength. What is still needed is evidence of the effect of silk on nest design and speciation after the point of its adoption within a family.

9.3 Taxonomic characters from nests

In the study of the evolution of nest design in the Hirundinidae (Winkler & Sheldon 1993), the phylogeny of the group was established from the similarity of DNA sequences. Traditionally, of course, phylogenies have been based upon anatomical characters of the organisms themselves; however, in some analyses, behavioural characters have been used to establish degrees of relatedness. The ability to recognise homology between behaviour patterns has been a matter of debate as behavioural characters have been regarded by some as too labile, and hence complicated by 'homoplasies' (characters shared through independent evolution), to make good taxonomic characters. Atz (1970) outlines safeguards to overcome the problems of using behavioural characters, including the need to confine any study to a group of closely related species to provide more confidence of homology. De Queiroz and Wimberger (1993), comparing behavioural and morphological characters for their effectiveness in phylogenetic studies, demonstrated similar levels of homoplasy for both, as did Kennedy, Spencer and Gray (1996) in a study which compared a phylogenetic tree based upon the social displays of Pelicaniformes with those based on molecular and morphological characters.

Lee *et al.* (1996) tested the validity of behavioural characters preserved in nests in the establishment of the phylogeny of the Asian swiftlets (Apodidae). The characters were the use of salivary mucus as an attachment material, the presence of vegetation and of feathers in the nest, and the degree of nest aggregation. The swifts are the only bird family to have developed the use of salivary mucus as a building material and nest attachment adhesive. They are also a family that shows great versatility in nest design and construction in its 18 genera; there are minimal nests of a few feathers stuck down with mucus, substantial cups of vegetation plus mucus, a token cup of feathers attached to a vertical palm leaf to which eggs are stuck with mucus (*African palm-swift, *Cypsiurus parvus*), a roof-attached nest with a long entrance tube made of mucus, plant fibres and plant down (*lesser swallow-tailed swift, *Panyptila cayennensis*), and a rock-wall-attached cup made of mucus alone (edible-nest swiftlet, *Collocalia* (*Aerodramus*) *fuciphaga*) (Lack 1956, Haverschmidt & Mees 1994).

If these innovations have had important evolutionary consequences, then, as in the swallows and martins, nest characters might be valuable in constructing a phylogeny. The taxonomic relationships between Asian swiftlets have proved difficult to determine using traditional criteria; however, in the analysis of Lee *et al.* (1996), a phylogeny based on mitochondrial DNA was in broad agreement with accepted classification. On the other hand, for no nest character was the fit between its occurrence and the molecular phylogeny better than chance, and the conclusion was that the nest characters were not a reliable guide to phylogeny, at least for this family. Such a result should not, however, condemn all nest characters in all bird families. Indeed, Clayton and Harvey (1993) argue that, because of the tangible nature of nests, they provide a particularly suitable source of characters and that the pattern of evolution of nest building in the Hirundinidae (Winkler & Sheldon 1993) provides reassurance of their validity in phylogenetic analyses.

In the general absence of precise and systematic descriptions of nests of closely related species, it is at present necessary to look at the detailed evolutionary history of nest building by hanging such nest descriptions as we have upon phylogenies based on morphological or molecular characters. A particularly valuable opportunity for this is offered by the Tyrannidae, a family in which the nest building is varied and phylogenies at the generic level are available (Lanyon 1986, 1988a, 1988b, Prum & Lanyon 1989,

Mobley & Prum 1995). Fitting nest designs to these phylogenies reveals the evolution of nest diversity. From this, insights may be obtained into the nature of the links between nest change and speciation, and also the relative importance of technological and design innovations.

There is a different approach to the understanding of the evolution of nest building. That is to reveal through a comparative ecological study how selection pressures may affect nest design through evolution. This is the approach adopted by Crook (1960, 1963) in the study of the nests of African weavers (Ploceinae) and supplemented by his study of the relationship between their ecology, social behaviour, reproduction and feeding (Crook 1964). Crook (1963) constructs a scheme describing the radiation of ploceine nest design in rather broad outline based largely on nest characters. This may, indeed, group nests whose similarity is homoplasic; nevertheless, some key evolutionary changes in nest design and materials can be related to ecological differences, and the direction of evolutionary change inferred.

9.4 Variability and conservatism

Recognising the evolutionary consequences of nest innovation is a problem in the majority of bird families. In the Dacelonidae (kingfishers) there are species that excavate cavities in the ground, in terrestrial mud termite nests, in arboreal carton termite nests and in rotted tree trunks (Coates 1985), but the order in which these sites become adopted in evolution is unknown. Even greater problems arise in the interpretation of innovation where families show variability in site, materials and nest design. A number of families of passerines have species that nest in burrows or cavities as well as build dome or cup-shaped nests. The starlings (Sturnidae), for example, have cavity nesters (e.g. the *common starling, *Sturnus vulgaris*), species like the superb starling (*Lamprotornis (Spreo) superbus*) that builds a domed nest of thorn twigs supported from below (Van Someren 1956), and the *metallic starling (*Aplonis metallica*) with its hanging basket of interlocked vine tendrils.

Some families have nests of all three general categories, cavity (hole), open (cup) and dome, including a trio of important neotropical families, the Furnariidae, Tyrannidae and Thraupidae – Tanagers: treated as a family in Howard & Moore (1991), but as a tribe, Thraupini, in Sibley & Monroe (1990). Collias and Collias (1984) have argued that cavity nesting 'because it meets most of the requirements of a nest in terms of warmth and safety, tends to block further evolution of nests built of specific materials'. Collias and Collias (1984) and Collias (1997) rightly point out the prevalence of cavity nesting among passerine and non-passerine families. Forty-four per cent of above-ground-nesting passerine families have some hole-nesting species, but, by the same data, 48% of these families have species that construct domed nests and 63% have open nest-building species. These data support the view that most families show versatility rather than one design inhibiting subsequent evolutionary change. The virtual monopoly of cavity nesting in families of the Coraciiformes, Piciformes and Trogoniformes could be interpreted as an example of evolutionary constraint, i.e. that selection pressure favours a different nest design but this is prevented by the fitness penalty resulting in some other biological attribute. Constraint is a difficult condition to confirm (Maynard Smith *et al.* 1985); however, Ligon (1993) dismisses it as unlikely in the context of these cavity-nesting families, which show some evidence of present or recent alternative nesting styles. The benefits of cavity nesting will, in any case, vary with latitude, and the costs may differ greatly between species (Sections 2.6 and 6.7). The costs of construction will be high for woodpeckers (Picidae) that excavate cavities in sound wood, more modest for species such as trogons that peck or gouge rotted wood, and negligible for species that take over natural cavities. In circumstances where selection pressures favour constructed nests over cavity nesting, and in the absence of phylogenetic constraint, the evolution of constructed nests from cavity-nesting ancestors may be expected to occur. In parrots (Psittacidae) this appears to have taken place in the monk parakeet (*Myiopsitta monachus*).

The monk parakeet is a remarkable member of its family for building a massive, domed, communal, stick nest, yet all other 340 or so parrot species (Psittacidae) are cavity nesters (Forshaw 1989). Eberhard (1996) hypothesised that in monk parakeets nest construction evolved either from behaviour used to modify the nests of other species, as she observed them to do with the covered twig nests of the brown cacholote (*Pseudoseisura lophotes*), or from the elaboration of former cavity-lining behaviour.

Only five other species of parrots show any nest building behaviour, and all belong to the African genus *Agapornis*. They build their nests within an existing tree cavity, the rosy-faced lovebird (*A. roseicollis*) by constructing a cup, the other four by constructing a complete domed nest within the cavity. Three other species in the same genus show the typical parrot pattern of cavity nesting without any nest construction. On the basis of DNA sequencing,

Eberhard (1998) derived a phylogeny which showed that the five nest-building *Agapornis* were a monophyletic clade, diverging from the non-building species early in the evolution of the genus, and that the domed nest design was a subsequent elaboration of the cup. Eberhard (1998) argues that, in *Agapornis*, nest building has evolved among cavity nesters to enable them to extend the range of usable cavities. This now includes domed nests of other bird species, which are known to be adopted facultatively by four of the five nest building *Agapornis* species.

Bird families vary considerably in the range of nest designs they exhibit. The family Furnariidae (ovenbirds) is one that shows a high degree of nest variability, both in terms of materials and of architecture (Vaurie 1980, Fitzpatrick 1982). It is a family of about 280 species in 66 genera if the Dendrocolaptidae (woodcreepers) are included, as in Sibley and Monroe (1990). Members of this family are dull coloured and of rather small size, ranging from the larger *Furnarius* species at about 56 g, down to *Leptasthenura* species at 10 g (Dunning 1993). In a test of the validity of nest characters alone to resolve the phylogeny of nest architecture, Zyskowski and Prum (1999) subjected this problematic family to cladistic analysis based on 24 characters covering building materials, architectural features, and construction methods. Comparisons with the outgroups reveal that cavity nesting is plesiomorphic for the family; however, in the two lineages in which primitive cavity nesting has been lost, all species have evolved new structures to enclose the nest. In *Furnarius* it is a massive mud dome that creates the new enclosure, although some species in the genus show cavity nesting (Studer & Vielliard 1990).

The other solution to retaining the closed nest has been a roof of plant materials. This is expressed in a diverse clade of 27 genera. Within this major clade, six minor clades are at least partially resolved, including novel phylogenetic relationships based on nest character synapomorphies. For example, a clade of two genera shares the use of sphagnum moss as a building material, and a clade of three marsh-nesting genera shares a distinctive entrance design. The authors emphasise the need to seek additional evidence to confirm these indications; however, the findings indicate that the Furnariidae have been conservative in the requirement for an enclosed nest, yet innovative evolving solutions.

In the large family Corvidae (Sibley & Monroe 1990) it is the cup that could be said to be the characteristic design feature. If this family of 127 genera and 647 species is truly monophyletic, it represents a radiation in size embracing *Corvus* species of over 1000 g to *Myiagra* species at little more than 10 g (Dunning 1993). They collectively make use of all the main categories of building material, yet virtually all of them build a cup nest. Again, the typical pattern is not without exceptions. The black-billed magpie (*Pica pica*) may build a loose dome of twigs over the cup (Birkhead 1991), and the Tibetan ground jay (*Pseudopodoces humilis*) apparently excavates a burrow in earthen banks (Goodwin 1976).

The non-passerine Columbidae (pigeons and doves) is quite a large family of 310 species in 40 genera (Sibley & Monroe 1990). It shows even greater conservatism than the Corvidae in nest form, and uses a limited range of materials. Nest sites are varied: on the ground, on rock ledges, or placed in various sites in trees and bushes. Nevertheless, a flimsy platform of fine twigs is characteristic of all (Goodwin 1983). Yet the argument of constraint is again weak. The size of birds in this family is predominantly 150–500 g (Dunning 1993), so building with fine twigs would be fairly typical, and the family does show some innovation in the use of materials, notably in the adoption of vine tendrils to form a cradle of interlocking elements attached to fine branches by the wompoo fruit-dove (*Ptilinopus magnificus*).

It seems that, in general, evidence of phylogenetic constraint in the evolution of bird nests is unconvincing. Columbidae, Corvidae and Furnariidae all show a tendency to exhibit certain design features, but some species in all three families provide exceptions, demonstrating that, even if some constraints operate, at least some genotypes have been able to transcend them.

9.5 Weaver birds and the ecology of nest evolution

The nests of the African weavers are known in a considerable degree of detail (Crook 1960, 1963, Collias & Collias 1964b), and upon these characters Crook (1963) has produced a scheme illustrating evolutionary trends based upon nest characters. These evolutionary trends are interpreted primarily in terms of selection pressures from the physical environment, particularly the rainfall that is an important variable in their African habitats.

The two genera (three species) of buffalo weavers (*Bubalornis* and *Dinemellia*), which may be included in the sub-family Ploceinae (Sibley & Monroe 1990) or excluded (Howard & Moore 1991), build massive structures in trees composed of an outer shell of thorny twigs lined with a core of dry grass stems or other soft materials. They are rather large birds relative to the true weavers, weighing in the region of 65 g (Dunning, 1993). Members of the related genera *Ploceopasser* (40–45 g), *Pseudonigrita* (20 g) and the sociable weaver (*Philetarius socius*, 27 g), all included by Sibley and Monroe (1990) in the Ploceinae, although lacking the thorny twig outer covering, share the characters of the two other genera and of the non-weaving Passeridae, i.e. nests supported from below on several branches (bottom multiple, branched) and extensive use of dry grass, generally stems with seed heads. The principle of construction using these materials lacks knots, loops and hitches, but relies on a technique of pushing the grass stems into the nest wall to build up its solidity (Collias & Collias 1964b). The transition from nest builders like this to true weaving appears to have been made possible by a technological revolution, building with fresh, green plant materials, either grass leaf or strips of palm frond. These new materials were flexible and capable of vigorous manipulation without fracture; attachments could be made by binding plant strips round a stem, or a fabric created by threading or knotting (see Fig. 4.16). The new material also spawned new nest shapes, some of them suspended from attachments a few millimetres in diameter. These design innovations were created by birds in the size range 20–40 g, similar to their probable ancestors.

Crook (1963) groups the nests of these true weavers into four modes. Nests of Mode A exhibit a variety of designs of woven nest including those with a long, downward projecting entrance tube of fine grass or palm strands (e.g. the black-necked weaver, *Ploceus nigricollis*, and the *red-vented malimbe, *Malimbus scutatus*) (Fig. 9.2). Also in this group are the simple, domed, palm-strip basket nests of *Euplectes* species and the baglafecht weaver (*P. baglafecht*). Similar to *Euplectes* are the Mode D nests of the *grosbeak weaver (*Amblyospiza albifrons*) and the compact weaver (*Pachypantes superciliosus*). Whether nests with or without an entrance tube represent the ancestral form, both Mode A and D nests share the feature of an outer woven shell, probably adapted for dry habitats (Crook 1963). Some species, however, have developed specialised features for rain shedding such as incorporating mud into the nest lining and thickening the nest wall by adding vegetation to the outside, e.g. baya weaver (*Ploceus philippinus*) – Mode A.

Mode B nests are distinguished by an architectural innovation, the *middle roof layer*, an additional layer added inside the egg chamber by the male before the female completes the nest. Nests in this group are found in humid sites in the African savannah, the middle roof layer apparently being an adaptation for rain shedding

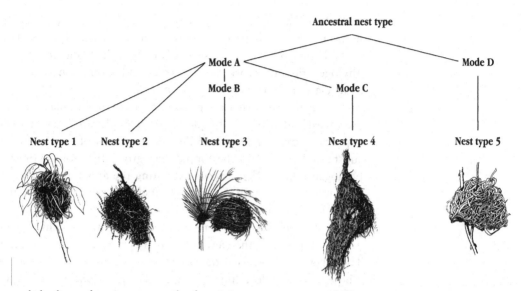

and the loss of an entrance tube for giving greater nest stability in the exposed windy sites. The remaining mode, Mode C, contains only one species, the *red-headed weaver (*Anaplectes* (*Malimbus*) *rubriceps*), which has produced yet another technological innovation, apparently abandoning palm or grass strips for green twigs torn at the base to produce a loose tag of bark which *in situ* can be used to bind the twig to others as it dries. Crook (1963) interprets this as a secondary adaptation to dry savannahs with an innovative solution to the problem of available flexible materials, but one apparently that has given rise to no further diversification.

Even lacking an independent and detailed cladogram for this group, it can still be concluded that the evolution of nest building has been characterised by changes in both design and in technology, and that these can in some cases be related to changes in selection pressure from physical features of the environment.

9.6 The Tyrannidae and the flexibility of building behaviour

In the Tyrannidae, the phylogenetic relationship between small groups of genera has been determined using a suite of characters, mainly anatomical features plus one or two from the nest (Lanyon 1986, 1988a, 1988b, Prum & Lanyon 1989, Mobley & Prum 1995). The family is valuable for the study of the evolution of nests because not only is it large (537 species in 146 genera (Sibley and Monroe 1990), but it is also very diverse in the nest designs exhibited within it. About 60% of genera build open cup nests, about

Figure 9.2
The scheme of Crook (1963) to explain the adaptive radiation of nest types in the weavers. Modes A and D are regarded as most closely approximating to the ancestral form. The nests in Mode A are more varied in form, some (type 1) without an entrance tube (e.g. baglafecht weaver, *Ploceus baglafecht*); some (type 2) with an entrance tube (e.g. lesser masked weaver, *P. intermedius*). *Amblyospiza albifrons (grosbeak weaver) (type 5) is one of two species assigned to Mode D. Mode B (e.g. northern brown-throated weaver, *P. castanops*) (type 3) is derived from Mode A, with the innovation of the middle roof layer. Mode C (e.g. *red-headed weaver, *Anaplectes rubriceps*) (type 4) results from a change in choice of materials from grass leaf to green twigs. (Adapted from Crook, 1963.)

30% domed nests, and a further 10% ground burrow or tree cavity nests (Traylor & Fitzpatrick 1982). By attaching descriptions of the nests based on the nest survey or published sources to the phylogenetic framework, the frequency and nature of nest changes as speciation occurred can be determined.

The results show that the possible types of evolutionary change are varied and also quite frequent. For example, an ancestral open cup design may be preserved but other aspects of the nest show substantial change. In the Empidonax group (Lanyon 1986), the phylogeny shows *Xenotriccus* exhibiting the apparently ancestral cup nest with bottom multiple (branched) support from which cavity nesting has evolved to be expressed in three extant genera (Fig. 9.3). The remaining genera in the cladogram all exhibit open nesting, supporting the view that this was the ancestral condition. These genera also reveal the origin of a new nest site preference where nests 'saddle' a limb or ledge (*Sayornis, Contopus*), and the adoption of a new material, mud (*Sayornis*) (Lanyon 1986). In two genera (*Mitrephanes, Empidonax*) a new nest attachment (top lip) has appeared to create a hammock design, although whether this arose once or more is unclear. So, in a monophyletic group which contains species of fairly similar size (6–20 g), nests, which can all be described as open in design, differ widely in materials, weight, attachment and site. This reveals the limitations of using broad nest categorisations such as open (cup), dome and hole. The nests of four genera in the Empidonax group could be classified as open, as if some common explanation could cover them all, but closer examination reveals interesting differences which have arisen through selection or drift as the group diversified.

Evolutionary changes from open to a domed design can be seen in some other groups of Tyrannidae. In the Schiffornis group of genera, the becards (*Pachyramphus*), with their characteristic globular nests either supported below or suspended from the top, have apparently arisen in a group of open nesters (Prum & Lanyon 1989). In the Elaenia group of 18 genera, despite some gaps in our knowledge of the nests, it is apparent that the domed nests of both *Tyranniscus* and *Camptostoma* evolved independently within a group of largely open cup nesters (Lanyon 1988a). Similarly, the loose, domed nests of *Myiobius*, as judged by their relationship to the open nest building *Myiophobus, Pyrrhomyias* and *Hirundinea*, have evolved from ancestors building open nests (Mobley & Prum 1995).

Evolution within the Phylloscartes group of nine genera and 42 species shows the emergence of a domed nest design from open nest

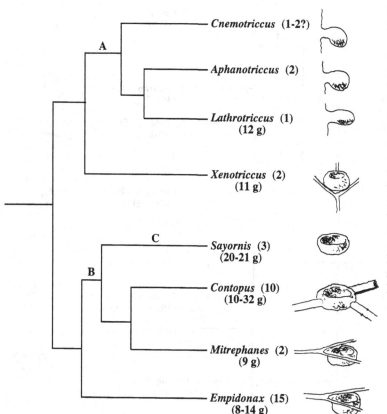

Figure 9.3
Phylogenetic relationships
of the Empidonax group
(Tyrannidae) (8 genera;
37 species) based largely
on anatomical characters.
The types of nest
associated with each
genus drawn from various
sources are illustrated.
A = nests in crevices or
tree cavities; B = nests
typically 'saddled' on a
limb or ledge; C = mud in
the nest foundation.
(Adapted from Lanyon
1986; bird weights from
Dunning 1993.) (Nest
sketches by Jane
Paterson.)

ancestors, within which has arisen a diversification of nest site choice and attachment, *Zimmerius* and *Phylloscartes* using bottom multiple support for nests sited in dense vegetation, while *Mionectes* and *Leptopogon* build suspended nests near water (Lanyon 1988a, Fig. 9.4). The cup nesters show diversity in nest attachment, with nests supported from below (*Sublegatus* and *Inezia*) or top lip with a hammock design (*Myiophobus* species).

In the Flatbill and Tody-tyrant assemblage of ten genera (65 species), a domed nest is apparently the ancestral condition (Lanyon 1988b). Its form is probably represented now by *Todirostrum*, *Poecilotriccus*, *Hemitriccus*, *Lophotriccus* and *Oncostoma*, all of which build an oval-shaped nest suspended at the top and with a side entrance (Fig. 9.5). From this, two distinct modifications of the domed form have arisen. In one, a porch has been added to the side of the nest to produce a retort design (*Rhynchocyclus*, *Tolmomyias*); in the other, the hanging nest has been greatly elongated by the construction of a *head* and loosely constructed hanging *tail* below the nest chamber to create nests over 1.5 m long, a design

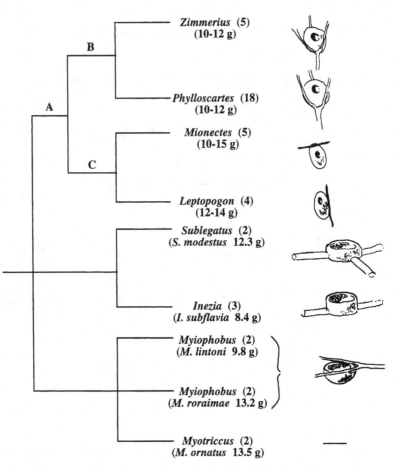

which, if the phylogeny is correct, arose in two genera independently (*Onychorhynchus* and *Cnipodectes*). To complete the virtuosity
of the assemblage, one genus (*Platyrinchus*) has evolved a cup-
shaped, bottom multiple-supported nest, described by Skutch
(1960) in the golden-crowned spadebill (*P. coronatus*) as resembling 'that of a hummingbird'.

Looking at nest evolution at the level of genus in the Tyrannidae
shows that the broad categories of open or domed nests are not very
helpful, because switches between these categories may not be
infrequent and various significant changes may occur within them.
There remains a substantial problem of interpretation. This cannot
be overcome without at least having more detailed descriptions of
the nests and of nest ecology (i.e. site, nest material availability,
predation, etc.). Detailed nest descriptions would also help to distinguish between changes which are purely of design and those
which are based on changes in technology. Examination of the role

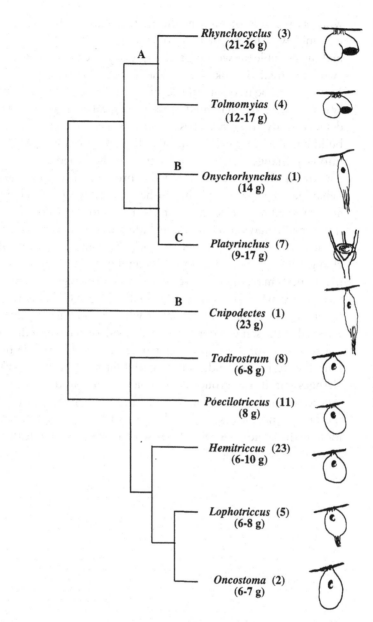

Figure 9.5
Phylogenetic relationships within the Flatbill and Tody-tyrant assemblage (10 genera; 65 species). Nest designs matched to this tree show closed, hanging nests of at least three distinct designs, with the genus *Platyrinchus* building an open cup. Within the phylogeny, a retort nest with downward directed tube has evolved (A), and a greatly elongated nest design has evolved, apparently on two separate occasions (B), and a cup nest (C) evolved from a covered nest ancestry. (Adapted from Lanyon 1988b, with the addition of bird weights from Dunning 1993, and nest sketches by Jane Paterson.)

of silk, in particular, in the changes observed in the Tyrannidae is required, as it is a material that can, for example, facilitate the transition from a nest supported from below and predominantly in compression to a suspended nest largely in tension.

The role of technological and design changes in the evolution of nests in some bird groups deserves more study. In swifts, a role for saliva in nest diversification is unsupported (Lee *et al.* 1996), whereas the role of mud in the nest evolution of swallows and martins (Winkler & Sheldon 1993) and of green plant strips in making possible the weaving technique of ploceine weaver birds are evidently great (Crook 1963). The Tyrannidae show that design changes may be frequent and accompanied by changes in nest site. Whether these are facilitated by technological changes awaits more detailed study. Together, these examples seem to confirm that nest building, like other behaviour, may be subject to quite rapid evolutionary change. This is consistent with the broad conclusion of Chapter 4 that nests are often the product of a rather limited behaviour repertoire and that rather similar building behaviours are observed in a wide range of species. Hansell (1984) describes how small behaviour changes, including changes in the behaviour sequences or loss of components from a sequence, can bring about changes in animal architecture. Changes of technology and design in bird nests may, therefore, result from small changes in behaviour, the former where an essentially unaltered behaviour is directed towards a novel material, the latter where a modified behaviour is directed at the same materials. Detailed video analysis of the building repertoire of a group of related species could provide behavioural criteria for understanding the history of nest building changes within the group. A suitable conclusion to this chapter, indeed to this book, is that, if we want to understand more about the importance of nests in the biology of birds, it is necessary to look at them more closely. What we already know indicates that this will be rewarding.

References

Addicott, A.B. (1938). Behaviour of the bush-tit in the breeding season. *Condor* 50, 49–63.

Aichorn, A. (1989). Nestbautechnik des Schneefinken (*Montifringilla n. nivalis* L.). *Egretta* 32, 58–71.

Alatalo, R.V., Carlson, A. & Lundberg, A. (1988). Nest cavity size and clutch size of pied flycatchers *Ficedula hypoleuca* breeding in natural tree holes. *Ornis Scandinavica* 19, 317–19.

Alcock, J. (1972). The evolution of the use of tools by feeding animals. *Evolution* 26, 464–73.

Alerstam, T. & Högstedt, G. (1981). Evolution of hole-nesting in birds. *Ornis Scandinavica* 12, 188–92.

Alvarez, L.W., Alvarez, W., Asaro, F. & Michel, H.V. (1980). Extraterrestrial cause for the Cretaceous–Tertiary extinction. *Science* 208, 1095–108.

Anderson, D.J., Stoyan, N.C. & Ricklefs, R.E. (1987). Why are there no viviparous birds? A comment. *American Naturalist* 130, 941–7.

Andersson, M. (1994). *Sexual Selection*. Princeton, NJ: Princeton University Press.

Andersson, S. (1989). Sexual selection and cues for female choice in leks of Jackson's widowbird *Euplectes jacksoni*. *Behavioural Ecology and Sociobiology* 25, 403–10.

Andersson, S. (1991). Bowers on the savannah: display courts and mate choice in a lekking widowbird. *Behavioural Ecology* 2, 210–18.

Andersson, S. (1992). Female preference for long tails in lekking Jackson's widowbirds: experimental evidence. *Animal Behaviour* 43, 379–88.

Andersson, M. & Wiklund, C.G. (1978). Clumping versus spacing out: experiments on nest predation in fieldfares (*Turdus pilaris*). *Animal Behaviour* 26, 1207–12.

Andrén, H. (1991). Predation: an overrated factor for over-dispersion of birds' nests? *Animal Behaviour* 41, 1063–9.

Andrews, R.C. (1932). *The New Conquest of Central Asia, a Narrative of the Explorations of the Central Asiatic Expeditions in Mongolia and China, 1921–30. Natural History of Central Asia,* Vol. 1. New York: American Museum of Natural History.

Apanius, V., Deerenberg, C., Daan, S. & Bos, N. (1996). Reproductive effort decreases antibody responsiveness. In *Parental Energy and Fitness Costs in Birds*, ed. C. Deerenberg, pp. 93–106, Wageningen: Posen en Looijen BV.

Arendt, W.J. (1985a). *Philornis* ectoparasitism of pearly-eyed thrashers. I. Impact on growth and development of nestlings. *Auk* 102, 270–80.

Arendt, W.J. (1985b). *Philornis* ectoparasitism of pearly-eyed thrashers. II. Effects on adults and reproduction. *Auk* 102, 281–92.

Armstrong, S. (1988). Pied a terre. *New Scientist* March 10th, 60–4.

Atz, J.W. (1970). The application of the idea of homology to behaviour. In *Development and Evolution of Behaviour: Essays in Honour of T.C. Schneirla*, ed. L.R. Aronson, E. Tobach, D.S. & Lehrman, J.S. Rosenblatt, pp. 53–74. San Francisco: W.H. Freeman.

Bailey, G.N. (1977). Shell mounds, shell middens and raised beaches in Cape York peninsula. *Mankind* 11, 132–43.

Balda, R.P. & Bateman, G.C. (1973). The breeding biology of the pinon jay. *Living Bird* 11, 5–42.

Barber, J.T., Ellgaard, E.G., Thien, L.B. & Stack, A.E. (1989). The use of tools for food transportation by the imported fire ant, *Solenopsis invicta*. *Animal Behaviour* 38, 550–2.

Barclay, R.M.R. (1988). Variation in the costs, benefits, and frequency of nest reuse by barn swallows (*Hirundo rustica*). *Auk* 105, 53–60.

Bateson, P. (1988). The active role of behaviour in evolution. In *Evolutionary Processes and Metaphors*, ed. M.W. Ho & S.W. Fox, pp. 191–207. Chichester: John Wiley.

Beck, B.B. (1980). *Animal Tool Behaviour*. New York: Garland STPM.

Béraut, E. (1970). The nesting of *Gymnoderus foetidus*. *Ibis* 112, 256.

Berenbaum, M.R., Green, E.S. & Zangerl, A.R. (1993). Web costs and web defence in the parsnip webworm (Lepidoptera, Oecophoridae). *Environmental Entomology* 22, 791–5.

Best, L.B. & Stauffer, D.F. (1980). Factors affecting nesting success in riparian bird communities. *Condor* 82, 149–58.

Birkhead, T.R. (1991). *The Magpies. The Ecology and Behaviour of Black-billed and Yellow-billed Magpies*. London: Poyser.

Blackburn, D.G. & Evans, H.E. (1986). Why are there no viviparous birds? *American Naturalist* 128, 165–90.

Blanco, G. & Tella, J.L. (1997). Protective association and breeding advantages of choughs nesting in lesser kestrel colonies. *Animal Behaviour* 54, 335–42.

Boesch, C. (1996). The question of culture. *Nature* 379, 207–8.

Bogliani, G., Sergio, F. & Tavecchia, G. (1999). Woodpigeons nesting in association with hobby falcons: advantages and choice rules. *Animal Behaviour* 57, 125–31.

Bogliani, G., Tiso, E. & Barbieri, F. (1992). Nesting association between the woodpigeon (*Columba palumbus*) and the hobby (*Falco subbuteo*). *Journal of Raptor Research* 26, 263–5.

Borgia, G. (1985a). Bower destruction and sexual competition in the satin bowerbird (*Ptilonorhynchus violaceus*). *Behavioural Ecology and Sociobiology* 18, 91–100.

Borgia, G. (1985b). Bower quality, number of decorations and mating success of male satin bowerbirds (*Ptilonorhynchus violaceus*): an experimental analysis. *Animal Behaviour* 33, 266–71.

Borgia, G. (1986). Sexual selection in bowerbirds. *Scientific American* 254, 70–9.

Borgia, G. (1993). The cost of display in the nonresource based mating system of the satin bowerbird. *American Naturalist* 44, 734–43.

Borgia, G. (1995a). Threat reduction as a cause of differences in bower architecture, bower decoration and male display in two closely related bowerbirds *Chlamydera nuchalis* and *C. maculata. Emu* 95, 1–12.

Borgia, G. (1995b). Complex male display and female choice in the spotted bowerbird: specialised functions for different bower decorations. *Animal Behaviour* 49, 1291–301.

Borgia, G. (1995c). Why do bowerbirds build bowers? *American Scientist* 83, 542–7.

Borgia, G. (1997). Comparative behavioural and biochemical studies of bowerbirds and the evolution of bower-building. In *Biodiversity, II. Understanding and Protecting our Biological Resources*, ed. M.L. Reaka-Kudla, D.E. Wilson & E.O. Wilson, pp. 263–76. Washington, DC: Joseph Henry Press.

Borgia, G. & Gore, M.A. (1986). Feather stealing in the satin bowerbird (*Ptilonorhynchus violaceus*): male competition and the quality of display. *Animal Behaviour* 34, 727–38.

Borgia, G., Kaatz, I.M. & Condit, R. (1987). Flower choice and bower decoration in the satin bowerbird *Ptilonorhynchus violaceus*: a test of hypotheses for the evolution of male display. *Animal Behaviour* 35, 1129–39.

Borgia, G. & Mueller, U. (1992). Bower destruction, decoration stealing and female choice in the spotted bowerbird *Chlamydera maculata. Emu* 92, 11–18.

Borgia, G., Pruett-Jones, S. & Pruett-Jones, M. (1985). The evolution of bower-building and the assessment of male quality. *Zeitschrift für Tierpsychologie* 67, 225–36.

Boswall, J. (1977). Tool-using by birds and related behaviour. *Avicultural Magazine* 83, 88–97.

Boswall, J. (1983). Tool-using and related behaviour in birds: more notes. *Avicultural Magazine* 89, 94–108.

Bowman, G.B. & Harris, L.D. (1980). Effect of spatial heterogeneity on ground-nesting depredation. *Journal of Wildlife Management* 44, 806–13.

Bradbury, J.W. (1981). The evolution of leks. In *Natural Selection and Social Behaviour: Research and New Theory*, ed. R.D. Alexander & D.W. Tinkle, pp. 138–69. New York: Chiron Press.

Brawn, J.D. & Balda, R.P. (1988). Population biology of cavity nesters in Northern Arizona: do nest sites limit breeding densities? *Condor* 90, 61–71.

Brown, C.R. (1994). Nest microclimate, egg temperature, egg water loss, and eggshell conductance in cape weavers *Ploceus capensis. Ostrich* 65, 26–31.

Brown, C.R. & Brown, M.B. (1986). Ectoparasitism as a cost of coloniality in cliff swallows (*Hirundo pyrrhonota*). *Ecology* 67, 1206–18.

Brown, C.R. & Brown, M.B. (1991). Selection of high quality host nests by parasitic cliff swallows. *Animal Behaviour* 41, 457–65.

Brown, J.L. (1974). Alternate routes to eusociality in jays – with a theory for the evolution of altruism and communal breeding. *American Zoologist* 14, 63–80.

Bryant, D.M. (1979). Reproductive costs in the house martin (*Delichon urbica*). *Journal of Animal Ecology* 48, 655–75.

Bucher, E.H. (1988). Do birds use biological control against nest parasites? *Parasitology Today* 4, 1–3.

Buhler, P. (1981). The functional anatomy of the avian jaw apparatus. In *Form and Function in Birds*, Vol. 2, ed. A.S. King & J. McLelland, pp. 439–68. London: Academic Press.

Burger, J. (1974). Breeding adaptations of Franklin's gull *Larus pipixcan* to a marsh habitat. *Animal Behaviour* 22, 521–67.

Butler, J.M. & Roper, T.J. (1996). Ectoparasites and sett use in European badgers. *Animal Behaviour* 52, 621–9.

Calder, W.A. (1971). Temperature relationships and nesting of the calliope hummingbird. *Condor* 73, 314–21.

Calder, W.A. (1973a). An estimate of the heat balance of a nesting hummingbird in a chilling climate. *Comparative Biochemistry and Physiology* 46, 291–300.

Calder, W.A. (1973b). Microhabitat selection during nesting of hummingbirds in the Rocky Mountains. *Ecology* 54, 127–34.

Calder, W.A. & King, J.R. (1974). Thermal and caloric relations of birds. In *Avian Biology*, Vol. 4, ed. D.S. Farner & J.R. King, pp. 259–413. New York: Academic Press.

Campbell, A.J. (1901). *Nests and Eggs of Australian Birds*. Sheffield: Private.

Carlquist, S. (1980). *Hawaii, a Natural History*, 2nd edn. Honolulu: Pacific Tropical Botanic Gardens.

Carrascal, L.M., Moreno, J. & Amat, J.A. (1995). Nest maintenance and stone theft in the chinstrap penguin (*Pygoscelis antarctica*). 2. Effects of breeding group size. *Polar Biology* 15, 541–5.

Carson, H.L. (1982). Speciation as a major reorganisation of phylogenetic balances. In *Mechanisms of Speciation*, ed. C. Barigozzi, pp. 411–33. New York: Alan R. Liss.

Chapin, J.P. (1953). The birds of the Belgian Congo, Part 3. *Bulletin of the American Museum of Natural History* 75A, 1–821.

Chapin, J.P. (1954). The birds of the Belgian Congo. Part 4. *Bulletin of the American Museum of Natural History* 75B, 1–846.

Charnov, E.L. & Krebs, J.R. (1974). On clutch-size and fitness. *Ibis* 116, 217–19.

Chilton, J. & Choo, B.S. (1992). Reciprocal frame, long span structures. *Proceedings of the International Association of Shell and Spatial Structures. Canadian Society of Civil Engineers Congress*, 2, 100–9.

Chilton, J., Choo, B.S. & Popovic, O. (1994). Morphology of some three-dimensional beam grillage structures in architecture and nature. *Natuerliche Konstruktionen 9, Evolution of Natural Structures*, 3rd International Symposium of Sonderforchungsbereich 230, Stuttart, 19–24.

Chisholm, A.H. (1952). Bird–insect nesting associations in Australia. *Ibis* 94, 395–405.

Christe, P., Oppliger, A. & Richner, H. (1994). Ectoparasite affects choice and use of roost sites in the great tit, *Parus major. Animal Behaviour* 47, 895–8.

Clark, L. (1990). Starlings as herbalists: countering parasites and pathogens. *Parasitology Today* 6, 358–60.

Clark, L. & Mason, J.R. (1985). Use of nest material as insecticidal and anti-pathogenic agents by the European starling. *Oecologia* 67, 169–76.

Clark, L. & Mason, J.R. (1987). Olfactory discrimination of plant volatiles by the European starling. *Animal Behaviour* 35, 227–35.

Clark, L. & Mason, J.R. (1988). Effect of biologically active plants used as nest material and the derived benefit to starling nestlings. *Oecologia* 77, 174–80.

Clark, L. & Smeraski, C.A. (1990). Seasonal shifts in odor acuity by starlings. *Journal of Experimental Zoology* 255, 22–9.

Clayton, D.H. & Harvey, P.H. (1993). Hanging nests on a phylogenetic tree. *Current Biology* 3, 882–3.

Coates, B.J. (1985). *The Birds of Papua New Guinea*, Vol. 1. Alderly, Australia: Dove Publications.

Coates, B.J. (1990). *The Birds of Papua New Guinea*, Vol. 2. Alderly, Australia: Dove Publications.

Collias, E.C. & Collias, N.E. (1964a). The development of nest building in a weaverbird. *Auk* 81, 42–52.

Collias, E.C. & Collias, N.E. (1973). Further studies on development of nest building behaviour in a weaverbird (*Ploceus cucullatus*). *Animal Behaviour* 21, 371–82.

Collias, N.E. (1986). Engineering aspects of nest building in birds. *Endeavour* 10, 9–16.

Collias, N.E. (1997). On the origin and evolution of nest building by passerine birds. *Condor* 99, 253–70.

Collias, N.E. & Collias, E.C. (1962). An experimental study of the mechanisms of nest building in a weaverbird. *Auk* 79, 568–95.

Collias, N.E. & Collias, E.C. (1964b). The evolution of nest building in weaverbirds (Ploceidae). *University of California Publications in Zoology* 73, 1–162.

Collias, N.E. & Collias E.C. (1984). *Nest Building and Bird Behavior*. Princeton, NJ: Princeton University Press.

Collias, N.E. & Victoria, J.K. (1978). Nest and mate selection in the village weaverbird (*Ploceus cucullatus*). *Animal Behaviour* 26, 470–9.

Collis, K. & Borgia, G. (1993). The costs of male display and delayed plumage maturation in the satin bowerbird (*Ptilonorhynchus violaceus*). *Ethology* 94, 59–71.

Conrad, K.F. & Robertson, R.J. (1993) Clutch size in eastern phoebes (*Sayornis phoebe*). I. The cost of nest building. *Canadian Journal of Zoology* 71, 1003–7.

Contino, F. (1968). Observations on the nesting of *Sporophila obscura* in association with wasps. *Auk* 85, 137–8.

Cooper, A. & Penny, D. (1997). Mass survival of birds across the Cretaceous–Tertiary boundary: molecular evidence. *Science* 275, 1109–13.

Cooper, W.T. & Forshaw, J.M. (1977). *The Birds of Paradise and Bowerbirds*. Sydney: Collins.

Cott, H.B. (1940). *Adaptive Colouration in Animals*. London:, Methuen.

Cowles, R.B. (1930). The life history of *Scopus umbretta bannermanii* C. Grant in Natal, South Africa. *Auk* 2, 159–76.

Cox, G.W. (1984). The distribution and origins of mima mound grasslands in San Diego County, California. *Ecology* 65, 1397–405.

Cox, G.W. & Roig, V.G. (1986). Argentinean mima mounds occupied by ctenomid rodents. *Journal of Mammalogy* 67, 428–32.

Cresswell, W. (1997). Nest predation: the relative effects of nest characteristics, clutch size and parental behaviour. *Animal Behaviour* 53, 93–103.

Cronin, H. (1991). *The Ant and the Peacock*. Cambridge: Cambridge University Press.

Crook, J.H. (1960). Nest form and construction in certain African weaverbirds. *Ibis* 102, 1–25.

Crook, J.H. (1963). A comparative analysis of nest structure in the weaver birds (Ploceinae). *Ibis* 105, 238–62.

Crook, J.H. (1964). The evolution of social organisation and visual communication in weaverbirds Ploceinae. *Behaviour* (Suppl.) 10, 1–178.

Darlington, J.P.E.C. (1984). Two types of mound built by the termite *Macrotermes subhyalinus* in Kenya. *Insect Science and its Application* 5, 481–92.

Darwin, C. (1859). *On the Origin of Species*. London: John Murray.

Darwin, C. (1871). *The Descent of Man* (2nd edition 1922). John Murray: London.

Davidson, G.W.H. (1981). Sexual selection and the mating system of *Argusianus orgus* (Aves: Phasianidae). *Biological Journal of the Linnaean Society* 15, 91–104.

Davidson, G.W.H. (1982). Sexual displays of the great argus pheasant *Argusianus orgus*. *Zeitschrift für Tierpsychologie* 58, 185–202.

Davies, S.J.J.F. (1962). Nest building of the magpie goose *Anseranas semipalmata*. *Ibis* 104, 147–57.

Dawkins, R. (1982). *The Extended Phenotype*. Oxford: Freeman.

Dawson, W.D., Lake, C.E. & Schumpert, S.S. (1988). Inheritance of burrow building in *Peromyscus*. *Behaviour Genetics* 18, 371–82.

Deeming, C. & Ferguson, M. (1989). In the heat of the nest. *New Scientist* 25th March, 33–8.

Delaney, M.F. & Linda, S.B. (1998). Characteristics of florida grasshopper sparrow nests. *Wilson Bulletin* 110, 136–9.

De Queiroz, A. & Wimberger, P.H. (1993). The usefulness of behaviour for phylogeny estimation: levels of homoplasy in behavioural and morphological characters. *Evolution* 47, 46–60.

Diamond, J.M. (1982a). Evolution of bowerbirds' bowers: animal origins of the aesthetic sense. *Nature* 297, 99–102.

Diamond, J.M. (1982b). Rediscovery of the yellow-fronted gardener bowerbird. *Science* 216, 431–4.

Diamond, J. (1986a). Animal art: variation in bower decorating style among male bowerbirds *Amblyornis inornatus*. *Proceedings of the National Academy of Sciences, USA* 83, 3042–6.

Diamond, J. (1986b). Biology of the birds of paradise and bowerbird. *Annual Review of Ecology and Systematics* 17, 17–37.

Diamond, J. (1987). Bower building and decoration by the bowerbird *Amblyornis inornatus*. *Ethology* 74, 117–204.

Diamond, J. (1988). Experimental study of bower decoration by the bowerbird *Amblyornis inornatus*, using colored poker chips. *American Naturalist* 131, 631–53.

Dilger, W.C. (1962). The behaviour of lovebirds. *Scientific American* 206 (1), 88–98.

Dobkin, D.S., Rich, A.C., Pretare, J.A. & Pyle, W.H. (1995). Nest-site relationships among cavity-nesting birds of riparian and snowpocket aspen woodlands in the Northeastern Great Basin. *Condor* 97, 694–707.

Dolnik, V.R. (1991). Time and energy needed for nest-building. *Soviet Journal of Zoology* 70, 97–106.

Dudgeon, D. (1987). A laboratory study of optimal behaviour and the costs of nest construction by *Polycentropus flavomaculatus* (Insecta: Trichoptera: Polycenropidae). *Journal of Zoology* 211, 121–41.

Duffy, D.C. (1983). The ecology of tick parasitism on densely nesting Peruvian seabirds. *Ecology* 64, 110–19.

Dunbrack, R.L. & Ramsay, M.A. (1989). The evolution of viviparity in amniote vertebrates: egg retention versus egg size reduction. *American Naturalist* 133, 138–48.

Dunning, J.B. (1993). *CRC Handbook of Avian Body Masses*. Boca Raton: CRC Press.

Eadie, J.McA., Kehoe, F.P. & Nudds, T.D. (1988). Pre-hatch and post-hatch brood amalgamation in North American Anatidae: a review of hypotheses. *Canadian Journal of Zoology* 66, 1709–21.

Eberhard, J. (1996). Nest adoption by monk parakeets. *Wilson Bulletin* 108, 374–7.

Eberhard, J.R. (1998). The evolution of nest-building behaviour in *Agapornis* parrots. *Auk* 115, 455–64.

Eberhard, W.G. (1988). Combing and sticky silk attachment behaviour by cribellate spiders and its taxonomic implications. *Bulletin of the British Arachnological Society* 7, 247–51.

Emlen, J.T. (1954). Territory, nest building, and pair formation in the cliff swallow. *Auk* 71, 16–35.

Endler, J.A. (1981). An overview of the relationship between mimicry and crypsis. *Biological Journal of the Linnaean Society* 16, 25–31.

Erikstad, K.E. & Tveraa, T. (1995). Does the cost of incubation set limits to clutch size in common eiders *Somateria mollissima*? *Oecologia* 103, 270–4.

Erpino, M.J. (1968). Nest-related activities of black-billed magpies. *Condor* 70, 154–65.

Evans, M.R. & Burn, J.L. (1996). An experimental analysis of mate choice in the wren: a monomorphic polygynous passerine. *Behavioural Ecology* 7, 101–8.

Evans, R.W. & Hatchwell, B.J. (1993). New slants on ornament asymmetry. *Proceedings of the Royal Society of London (B)* 251, 171–7.

Ewins, P.J., Miller, M.J.R., Barker, M.E. & Postpalsky, S. (1994). Birds breeding in or beneath osprey nests in the Great Lakes basin. *Wilson Bulletin* 106, 743–9.

Falla, R.A., Sibson, R.B. & Turbott, E.G. (1989). *Collins' Guide to the Birds of New Zealand and Outlying Islands.* Auckland: Collins.

Feduccia, A. (1995). Explosive evolution in Tertiary birds and mammals. *Science* 267, 637–8.

Felsenstein, J. (1985). Phylogenies and the comparative method. *American Naturalist* 125, 1–15.

Fenske-Crawford, T.J. & Neimi, G.J. (1997). Predation of artificial ground nests at two types of edges in a forest-dominated landscape. *Condor* 99, 14–24.

ffrench, R. (1991). *A Guide to the Birds of Trinidad and Tobago,* 2nd edn. Ithaca: Cornell University Press.

Fisher, R.A. (1930). *The Genetical Theory of Natural Selection.* Oxford: Clarendon Press.

Fitzpatrick, J.W. (1982). Taxonomy and geographical distribution of the Furnariidae (Aves, Passeriformes). *Auk* 99, 810–13.

Ford, M.J. (1977). Energy costs of predation strategy of web-spinning spider *Lepthyphantes zimmermanni* Bertkau (Linyphilidae). *Oecologia* 28, 341–9.

Forshaw, J.M. (1989). *Parrots of the World,* 3rd (revised) edn. Melbourne: Lansdowne.

Friedmann, H. (1922). The weaving of the red-billed weaver bird, *Quelea quelea*, in captivity. *Zoologia* 2, 355–72.

Frith, C.B., Borgia, G. & Frith, D.W. (1996). Courts and courtship behaviour of Archbold's bowerbird *Archboldia papuensis. Ibis* 138, 204–11.

Frith, C.B. & Frith, D.W. (1994). Courts and seasonal activities at them by male tooth-billed bowerbirds, *Scenopoeetes dentirostris* (Ptilonorhynchidae). *Memoires of the Queensland Museum* 37, 121–45.

Frith, C.B. & Frith, D.W. (1995). Court site constancy, dispersion, male survival and court ownership in the male tooth-billed bowerbird *Scenopoeetes dentirostris* (Ptilonorhynchidae). *Emu* 95, 84–98.

Fry, C.H. (1972). The social organisation of bee-eaters (Meropidae) and co-operative breeding in hot-climate birds. *Ibis* 114, 1–14.

Fuller, E. (1987). *Extinct Birds*. Harmondsworth: Viking/Rainbird.

Gails, F. & Metz, J.A.J. (1998). Why are there so many cichlid species? *Trends in Ecology and Evolution* 13, 1–2.

Galbraith, H. (1988). Effects of agriculture on the breeding ecology of the lapwing *Vanellus vanellus*. *Journal of Applied Ecology* 25, 487–503.

Garson, P.J. (1980). Male behaviour and female choice: mate selection in the wren ? *Animal Behaviour* 28, 491–502.

Gehlbach, F.R. & Baldridge, R.S. (1987). Live blind snakes (*Leptotyphlops dulcis*) in eastern screech owl (*Otus asio*) nests: a novel commensalism. *Oecologia* 71, 560–3.

Gilliard, E.T. (1962). On the breeding behaviour of the cock-of-the-rock (Aves, *Rupicola rupicola*). *Bulletin of the American Museum of Natural History* 124, 35–58.

Gilliard, E.T. (1963). The evolution of bowerbirds. *Scientific American* 209, 38–46.

Gilliard, E.T. (1969). *Birds of Paradise*. New York: Natural History Press.

Glanville, R.R. (1954). *Picathartes gymnocephalus* in Sierra Leone. *Ibis* 96, 481–4.

Goodwin, D. (1976). *Crows of the World*. London: British Museum (Natural History) Publication No. 771.

Goodwin, D. (1983). *Pigeons and Doves of the World*. Ithaca, NY: British Museum (Natural History) Comstock Publishing Associates, Cornell University Press.

Gordon, J.E. (1978). *Structures*. Harmondsworth & New York: Penguin.

Gould, J. (1841). *The Birds of Australia (1841)*. London: Methuen.

Grassé, P.P. (1951). *Traité de Zoologie*, Vol. 10. Paris: Masson.

Greer, A.E. (1970). Evolutionary and systematic significance of crocodilian nesting habits. *Nature* 227, 523–4.

Greer, A.E. (1971). Crocodilian nesting habits and evolution. *Fauna (Rancho Mirage, California)* 2, 20–8.

Griffin, D.R. (1981). *The Question of Animal Awareness*. New York: The Rockefeller University Press.

Grigg, G.C. (1973). Some consequences of the shape and orientation of magnetic termite mounds. *Journal of Australian Zoology* 21, 231–7.

Grimes, L. & Darku, K. (1968). Some recent breeding records of *Picathartes gymnocephalus* in Ghana and notes on its distribution in West Africa. *Ibis* 110, 93–9.

Groom, M.J. (1992). Sand-colored nighthawks parasitize the antipredator behavior of three nesting bird species *Ecology* 73, 785–93.

Gustaffson, L. & Nilsson, S.G. (1985). Clutch size and breeding success of pied and collared flycatchers *Ficedula* spp. in nest boxes of different sizes. *Ibis* 127, 380–5.

Gustaffson, L. & Sutherland, W.J. (1988). The costs of reproduction in the collared flycatcher *Ficedula albicollis*. *Nature, London* 335, 813–15.

Gwinner, H. (1997). The function of green plants in nests of European starlings (*Sturnus vulgaris*). *Behaviour* 134, 337–51.

Haemig, P.D. (1999). Predation risk alters interactions among species: competition and facilitation between ants and nesting birds in a boreal forest. *Ecology Letters* 2, 178–84.

Hails, C.J. & Turner, A.K. (1984). The breeding biology of the Asian palm swift *Cypsiurus balasiensis*. *Ibis* 126, 74–81.

Halliday, T. (1978). Sexual selection and mate choice. In *Behavioural Ecology: an Evolutionary Approach*, ed. J.R. Krebs & N.B. Davies, pp. 180–213. Sunderland, MA: Sinaur.

Hamilton, W.D. (1964). The genetical theory of social behaviour, I, II. *Journal of Theoretical Biology* 7, 1–52.

Hamilton, W. & Zuk, M. (1982). Heritable true fitness and bright birds, a role for parasites? *Science* 218, 384–7.

Hansell, M.H. (1972). Case building behaviour of the caddis larva *Lepidostoma hirtum*. *Journal of Zoology, London* 167, 179–92.

Hansell, M.H. (1981). Nest construction in the subsocial wasp *Parischnogaster mellyi* (Saussure), Stenogastrinae. *Insectes Sociaux* 28, 208–16.

Hansell, M.H. (1984). *Animal Architecture and Building Behaviour*. London: Longman.

Hansell, M.H. (1986). Elements of eusociality in colonies of *Eustenogaster calyptodoma* (Sakagami and Yoshikawa) (Stenogastrinae, Vespidae). *Animal Behaviour* 35, 131–41.

Hansell, M.H. (1987a). Nest building as a facilitating and limiting factor in the evolution of eusociality in the Hymenoptera. In *Oxford Surveys of Evolutionary Biology*, Vol. 4, ed. P. Harvey & L. Partridge, pp. 155–81. Oxford: Oxford University Press.

Hansell, M.H. (1987b). What's so special about tool use? *New Scientist* January 8th, 54–6.

Hansell, M.H. (1993a). The ecological impact of animal nests and burrows. *Functional Ecology* 7, 5–12.

Hansell, M.H. (1993b). The long-tailed tit's nest of many parts. *BTO News, British Trust for Ornithology* 186, 20.

Hansell, M.H. (1995). The demand for feathers as a nesting material by woodland nesting birds. *Bird Study* 42, 240–5.

Hansell, M.H. (1996a). Wasps make nests: nests make conditions. In *Natural History and Evolution of Paper Wasps*, ed. S. Turillazzi & M.J. West-Eberhard, pp. 272–89. Oxford: Oxford University Press.

Hansell, M.H. (1996b). The function of lichen flakes and white spider cocoons on the outer surface of birds' nests. *Journal of Natural History* 30, 303–11.

Hansell, M.H. (in preparation). Specialisation in the materials and construction of the nest of the long-tailed tit (*Aegithalos caudatus*).

Hansell, M.H. & Aitken, J.J. (1977). *Experimental Animal Behaviour.* Glasgow: Blackie.

Hardy, J.W. (1963). Epigamic and reproductive behaviour of the orange-fronted parakeet. *Condor* 65, 169–99.

Harrison, C. (1975). *A Field Guide to the Nests, Eggs and Nestlings of British and European Birds.* London: Collins.

Harrison, C. (1978). *A Field Guide to the Nests, Eggs and Nestlings of North American Birds.* London: Collins.

Harrison, H.H. (1979). *A Field Guide to Western Birds' Nests.* Boston: Houghton Mifflin.

Harvey, P.H. & Purvis, A. (1991). Comparative methods for explaining adaptations. *Nature* 351, 619–24.

Haskell, D.G. (1996). Do bright colours at nests incur a cost due to predation? *Evolutionary Ecology* 10, 285–8.

Haverschmidt, F. (1957). Nachbarcshaft von vogelnestern und wespennestern Surinam *Journal für Ornithologie* 98, 389–96.

Haverschmidt, F. & Mees, G.F. (1994). *Birds of Suriname.* Paramaribo: Vaco.

Heaney, V. & Monaghan, P. (1995). A within-clutch trade-off between egg production and rearing in birds. *Proceedings of the Royal Society of London B* 261, 361–5.

Heath, M. & Hansell, M.H. (2000). Weaving techniques in two species of Icterini, the yellow oriole and collared oropendola. In *Studies in Trinidad and Tobago: Ornithology Honouring Richard ffrench*, ed. F. Hayes. Occasional papers of the Department of Life Sciences. St Augustine, Trinidad: University of the West Indies.

Hedger, J. (1990). Fungi in the tropical canopy. *Tropical Mycology News* 4, 200–2.

Hedges, S.B., Parker, P.H., Sibley, C.G. & Kumar, S. (1996). Continental break-up and the ordinal diversification of birds and mammals. *Nature (London)* 381, 226–9.

Henschel, J.R., Mendelsohn, J.M. & Simmons, R. (1991). Is the association between the Gabar goshawks and social spiders (*Stegodyphus*) mutalism or theft? *Gabar* 6, 57–60.

Henschel, J.R., Simmons, R.E. & Mendelsohn, J.M. (1992). Gabar goshawks and social spiders revisited: untangling the web. *Gabar* 7, 49.

Herrick, F.H. (1911). Nests and nest building in birds: Part 1. *Journal of Animal Behaviour* 1, 157–92.

Higuchi, H. (1986). Bait fishing by the green-backed heron *Ardeola striata* in Japan. *Ibis* 128, 286–90.

Hilton, G., Hansell, M.H., Ruxton, G.D. & Monaghan, P. (in preparation). Insulation matters: the effects of nesting material on heat loss from avian clutches.

Hindwood, K.A. (1959). The nesting of birds in the nests of social insects. *Emu* 59, 1–36.

Höglund, J. (1989). Size and plumage dimorphism in lek-breeding birds: a comparative analysis. *American Naturalist* 134, 72–87.

Högstedt, G. (1980). Evolution of clutch size in birds: adaptive variation in relation to territory quality. *Science* 210, 1148–50.

Hoi, H., Schleicher, B. & Valera, F. (1994). Female mate choice and nest desertion in penduline tits, *Remiz pendulinus*: the importance of nest quality. *Animal Behaviour* 48, 743–6.

Hölldobler, B. & Wilson, E.O. (1990). *The Ants*. Berlin: Springer-Verlag.

Hoogland, J.L. & Sherman, P.W. (1976). Advantages and disadvantages of bank swallow (*Riparia riparia*) coloniality. *Ecological Monographs* 46, 33–58.

Horner, J.R. (1982). Evidence of colonial nesting and 'site fidelity' among ornithiscian dinosaurs. *Nature* 297, 675–6.

Horner, J.R. (1984). The nesting behaviour of dinosaurs. *Scientific American* 240, 130–7.

Horner, J.R. & Weishampel, D.B. (1988). A comparative embryological study of two ornithiscian dinosaurs. *Nature* 332, 256–7.

Horváth, O. (1964). Seasonal differences in rufous hummingbird nest height and their relation to nest climate. *Ecology* 45, 235–41.

Hotta, M. (1994). Infanticide in little swifts taking over costly nests. *Animal Behaviour* 47, 491–3.

Howard, J. (1997). Sticky solutions. *New Scientist, Inside Science* No. 102, 1–4.

Howard, R. & Moore, A. (1991) *A Complete Checklist of Birds of the World*. London: Academic Press.

Howlett, J.S. & Stutchbury, B.J. (1996). Nest concealment and predation in hooded warblers: experimental removal of nest cover. *Auk* 113, 1–9.

Howman, H.R.G. & Begg, G.W. (1983). Nest building and nest destruction by the masked weaver, *Ploceus velatus*. *South African Journal of Zoology* 18, 37–44.

Howman, H.R.G. & Begg, G.W. (1995). Intra-seasonal and inter-seasonal nest renovation in the masked weaver, *Ploceus velatus*. *Ostrich* 66, 122–8.

Hrdy, S.B. (1979). Infanticide among animals: a review, classification and examination of the implications for the reproductive strategies of females. *Ethological Sociobiology* 1, 13–40.

Humphrey, P.S. & Peterson, R.T. (1978). Nesting behaviour and affinities of monk parakeets of southern Buenos Aires province, Argentina. *Wilson Bulletin* 90, 544–52.

Humphries, S. & Ruxton, G.D. (1999). Bower-building: coevolution of display traits in response to the costs of female choice? *Ecology Letters* 2, 404–13.

Hunt, G.R. (1996). The manufacture of hook-tools by New Caledonian crows. *Nature* 379, 249–51.

Hunter, C.P. & Dwyer, P.D. (1997). The value of objects to satin bowerbirds *Ptilonorhynchus violaceus*. *Emu* 97, 200–6.

Ito, Y. (1993). *Behaviour and Social Evolution of Wasps. The Communal Aggregation Hypothesis*. Oxford: Oxford University Press.

Jackson, J.A. & Burchfield, P.G. (1975). Nest-site selection of barn swallows in east-central Mississippi. *American Midland Naturalist* 94, 503–9.

Jacobs, C.H., Collias, N.E. & Fujimoto, J.T. (1978). Nest colour as a factor in nest selection by male village weaverbirds. *Animal Behaviour* 26, 463–9.

Jacobs, M.I. (1996). Working knowledge: unzipping Velcro. *Scientific American* 274 (4), 96.

Janzen, D.H. (1969). Birds and the ant × acacia interaction in Central America, with notes on birds and other myrmecophytes. *Condor* 71, 240–56.

Jeanne, R.L. (1975). The adaptiveness of social wasp nest architecture. *Quarterly Review of Biology* 50, 267–87.

Jeanne, R.L. (1978). Intraspecific nesting associations in the neotropical social wasp *Polybia rejecta* (Hymenoptera: Vespidae). *Biotropica* 10, 234–5.

Johnson, L.S. & Albrecht, D.J. (1993). Effects of haematophagous ectoparasites on nestling house wrens, *Troglodytes aedon*: who pays the cost of parasitism? *Oikos* 66, 255–62.

Johnston, R.F. & Hardy, J.W. (1962). Behaviour of the purple martin. *Wilson Bulletin* 74, 243–62.

Joyce, F.J. (1993). Nesting success of rufous-naped wrens (*Campylorhynchus rufinucha*) is greater near wasp nests. *Behaviour Ecology and Sociobiology* 32, 71–7.

Kahl, M.P. (1967). Observations on the behaviour of the hamerkop *Scopus umbretta* in Uganda. *Ibis* 109, 25–32.

Kang, N., Hails, C.J. & Sigurdsson, J.B. (1991). Nest construction and egg laying in edible-nest swiftlets *Aerodramus* spp. and the implications for harvesting. *Ibis* 133, 170–7.

Kear, J. (1970). Adaptive radiation of parental care in waterfowl. In *Social Behaviour in Birds and Mammals*, ed. J.H. Crook, pp. 357–93. London: Academic Press.

Kennedy, M., Spencer, H.G. & Gray, R.D. (1996). Hop, step and gape: do the social displays of the Pelicaniformes reflect phylogeny? *Animal Behaviour* 51, 273–91.

Kilgore, D.L. & Knudsen, K.L. (1977). Analysis of materials in cliff and barn swallow nests. Relationships between mud selection and nest architecture. *Wilson Bulletin* 89, 562–71.

Kirkpatrick, M. (1982). Sexual selection and the evolution of female choice. *Evolution* 36, 1–12.

Klaas, E.E. (1970). A population study of the eastern phoebe, *Sayornis phoebe*, and its social relationships with the brown-headed cowbird, *Molothrus ater*. PhD Thesis, Department of Zoology, University of Kansas, Laurence.

Koepcke, M. (1972). Über die Resistenzformen der Vogelnester in einem begrenzten Gebeit des tropischen Regenwaldes in Peru. *Journal für Ornithologie* 113, 138–60.

Krafft, B. (1966). Étude du comportement social de l'araignée (*Agelena consociata* Denis). *Biologia Gabonica,* 2, 235–50.

Krebs, J.R. (1990). Food storing birds: adaptive specialisation in brain and behaviour. *Philosophical Transactions of the Royal Society,* B 329, 153–60.

Kruuk, H. (1964). Predators and anti-predator behaviour of the black-headed gull (*Larus ridibundus* L.). *Behaviour Supplement* 11, 1–130.

Kulczycki, A. (1973). Nesting of the members of the corvidae in Poland. *Acta Zoologica Cracoviensia* 18, 583–666.

Kulesza, G. (1990). An analysis of clutch-size in New World passerine birds. *Ibis* 132, 407–22.

Kusmierski, R., Borgia, G., Crozier, R.H. & Chan, B.H.Y. (1993). Molecular information on bowerbird phylogeny and the evolution of exaggerated male characteristics. *Journal of Evolutionary Biology* 6, 737–52.

Kusmierski, R., Borgia, G., Uy, A. & Crozier, R.H. (1997). Labile evolution of display traits in bowerbirds indicates reduced effects of phylogenetic constraint. *Proceedings of the Royal Society of London,* B 264, 307–13.

Lack, D. (1947). The significance of clutch size. *Ibis* 89, 302–52.

Lack, D. (1956). *Swifts in a Tower.* London: Methuen.

Lande, R. (1981). Models of speciation by sexual selection on polygenic traits. *Proceedings of the National Academy of Sciences, USA* 78, 3721–62.

Lanyon, W.E. (1986). A phylogeny of thirty-three genera in the *Empidonax* assemblage of tyrant flycatchers. *American Museum Novitates* 2846, 1–64.

Lanyon, W.E. (1988a). A phylogeny of the thirty-two genera in the *Elaenia* assemblage of tyrant flycatchers. *American Museum Novitates* 2914, 1–57.

Lanyon, W.E. (1988b). A phylogeny of the flatbill and tody-tyrant assemblage of tyrant-flycatchers. *American Museum Novitates* 2923, 1–41.

Leader, N. & Yom-Tov, Y. (1998). Possible function of stone ramparts at the nest entrance of the blackstart. *Animal Behaviour* 56, 207–17.

Lee, P.L., Clayton, D.H., Griffiths, R. & Page, R.D.M. (1996). Does behaviour reflect phylogeny in the swiftlets (Aves: Apodidae)? A test using cytochrome *b* mitochondrial DNA sequences. *Proceedings of the National Academy of Sciences, USA* 93, 7091–6.

Lens, L. & Dhondt, A.A. (1993). Effects of habitat fragmentation on the timing of crested tit *Parus cristatus* natal dispersal. *Ibis* 136, 147–52.

Lens, L., Wauters, L.A. & Dhondt, A.A. (1994). Nest-building by crested tit *Parus cristatus* males: an analysis of costs and benefits. *Behavioural Ecology and Sociobiology* 35, 431–6.

Lenz, N. (1994). Mating behaviour and sexual competition in the regent bowerbird *Sericulus chrysocephalus. Emu* 94, 263–72.

Li, P. & Martin, T.E. (1991). Nest-site selection and nesting success of cavity-nesting birds in high elevation forest drainages. *Auk* 108, 405–18.

Ligon, J.D. (1970). Behaviour and breeding biology of the red-cockaded woodpecker. *Auk* 87, 255–78.

Ligon, J.D. (1993). The role of phylogenetic history in the evolution of contemporary avian mating and parental care systems. In *Current Ornithology*, Vol. 10, ed. D.M. Power, pp. 1–40. New York: Plenum Press.

Ligon, J.D. (1999). *The Evolution of Avian Breeding Systems*. Oxford: Oxford University Press.

Lill, A. (1974). The evolution of clutch size and male 'chauvinism' in the white-bearded manakin. *The Living Bird* 13, 211–31.

Lin, N. & Michener, C.D. (1972). Evolution of sociality in insects. *Quarterly Review of Biology* 47, 131–59.

Lindell, C. (1996). Patterns of nest usurpation: when should species converge on nest niches? *Condor* 98, 464–73.

Lindén, M. & Møller, A.P. (1989). Cost of reproduction and covariation of life history traits in birds. *Trends in Ecology and Evolution* 4, 367–71.

Linsenmaier, W. (1979). *Wonders of Nature*. New York: Random House.

Liversidge, R. (1962). The breeding biology of the little sparrowhawk *Accipiter minullus. Ibis* 104, 399–406.

Liversidge, R. (1963). The nesting of the hamerkop *Scopus umbretta. Ostrich* 34, 55–62.

Lockley, R.M. (1942). *Shearwaters*. London: J.M. Dent.

Lombardo, M.P. (1994). Nest architecture and reproductive performance in tree swallows (*Tachycineta bicolor*). *Auk* 111, 814–24.

Lombardo, M.P., Bosman, R.M., Faro, C.A., Houtteman, S.G. & Klusiza, T.S. (1995). Effect of feathers as nest insulation on incubation behaviour and reproductive performance of tree swallows (*Tachycineta bicolor*). *Auk* 112, 973–81.

Lounibos, L.P. (1975). The cocoon spinning behaviour of the Chinese oak silk moth *Antheraea pernyi. Animal Behaviour* 23, 843–53.

Lovegrove, B.G. (1991). Mima-like mounds (heuwelties) of South Africa: the topographical, ecological and economic impact of burrowing animals. In *The Environmental Impact of Burrowing Animals and*

Animal Behaviour, ed. P.S. Meadows & A. Meadows, pp. 183–98. Symposia of the Zoological Society of London, No. 63.

Luscher, M. (1961). Air conditioned termite nests. *Scientific American* 205, 138–45.

MacArthur, R.H. (1958). Population ecology of some warblers of north-eastern coniferous forests. *Ecology* 39, 599–619.

MacArthur, R.H., MacArthur, J.W. & Preer, J. (1962). On bird diversity II. Prediction of bird census from habitat measurements. *American Naturalist* 46, 167–74.

McCorquodale, D.B. (1989). Soil softness, nest initiation and nest sharing in the wasp *Cerceris antipodes* (Hymenoptera, Sphecidae). *Ecological Entomology* 14, 191–6.

McCrae, A.W.R. & Walsh, J.F. (1974). Association between nesting birds and polistine wasps in North Ghana. *Ibis* 116, 215–17.

McFarland, K.P. & Rimmer, C.C. (1996). Horsehair fungus *Marasmius androcaceus*, used as nest linings by birds of the sub-alpine spruce-fir community in North-eastern United States. *Canadian Field Naturalist* 110, 541–3.

McGrew, W.C. (1992). *Chimpanzee Material Culture: Implications for Human Evolution*. Cambridge: Cambridge University Press.

McKaye, K.R., Louda, S.M. & Stauffer, J.R. (1990). Bower size and male reproductive success in a cichlid fish lek. *American Naturalist* 135, 597–613.

Maclaren, P.I.R. (1950). Bird–ant nesting associations. *Ibis* 92, 564–6.

McNeil, D. & Clark, F. (1977). Nest architecture of house martins. *Bird Study* 24, 130–2.

Madge, S.G. (1970). Nest of the long-billed spiderhunter *Arachnothera robusta*. *Malay Naturalists Journal* 23, 125.

Marini, M.A. (1997). Predation-mediated bird nest diversity: an experimental test. *Canadian Journal of Zoology* 75, 317–23.

Marler, P. (1956). Behaviour of the chaffinch. *Behaviour*, Suppl. 5, 1–184.

Marshall, A.J. (1954). *Bowerbirds: Their Displays and Breeding Success*. Oxford: Clarendon Press.

Martella, M.B. & Bucher, E.H. (1984). Nesting of the spot-winged falconet in monk parakeet nests. *Auk* 101, 614–15.

Martin, T.E. (1988a). Nest placement: implications for selected life-history traits, with special reference to clutch size. *American Naturalist* 132, 900–10.

Martin, T.E. (1988b). Habitat and area of effects on forest bird assemblages: is nest predation an influence? *Ecology* 69, 74–84.

Martin, T.E. (1988c). On the advantage of being different: nest predation and the coexistence of bird species. *Proceedings of the National Academy of Science* 85, 2196–9.

Martin, T.E. (1992). Breeding productivity considerations: what are the appropriate habitat features for management? In *Ecology and Conservation of Neotropical Migrant Landbirds*, ed. J.M. Hagan III

& D.W. Johnston, pp. 455–71. Washington: Smithsonian Institution Press.

Martin, T.E. (1993a). Nest predation among vegetation layers and habitat types: revising the dogmas. *American Naturalist* 141, 897–913.

Martin, T.E. (1993b). Evolutionary determinants of clutch size in cavity-nesting birds: nest predation or limited breeding opportunities. *American Naturalist* 142, 937–46.

Martin, T.E. (1993c). Nest predation and nest sites. *Bioscience* 43, 523–32.

Martin, T.E. & Li, P. (1992). Life history traits of open vs. cavity nesting birds. *Ecology* 73, 579–92.

Martin, T.E. & Roper, J.J. (1988). Nest predation and nest-site selection of a western population of the hermit thrush. *Condor* 90, 51–7.

Mason, P. (1985). The nesting biology of some passerines of Buenos Aires, Argentina. In *Neotropical Ornithology. Ornithological Monographs No. 36*, ed. P.A. Buckley, M.S. Foster, E.S. Morton, R.S. Ridgely & F.G. Buckley, pp. 954–72. Washington, DC: American Ornithologists Union.

Matessi, G. & Bogliani, G. (1994). Experiments on nest predation: effect of habitat fragmentation and landscape features. *21st International Ornithological Congress, Vienna, Ornithological Notebook*, p. 647.

Matsuzawa, T. (1991). Nesting cups and metatools in chimpanzees. *Behavioural and Brain Sciences* 14, 570–2.

Maynard Smith, J., Burian, R., Kauffman, S. *et al.* (1985). Developmental constraints and evolution. *Quarterly Review of Biology* 60, 265–87.

Medway, Lord (1962). The swiftlets (*Collocalia*) of the Niah Cave, Sarawak. *Ibis* 104, 45–66.

Melin, E., Wikén, T. & Öblom, K. (1947). Antibiotic agents in the substrates from cultures of the genus *Marasmius*. *Nature* 159, 840–1.

Merino, S. & Potti, J. (1995). Pied flycatchers prefer to nest in clean nest boxes in an area with detrimental nest ectoparasites. *Condor* 97, 828–31.

Merton, D.V., Morris, R.B. & Atkinson, I.A. (1984). Lek behaviour in a parrot: the kakapo (*Strigops hapbroptilus*) of New Zealand. *Ibis* 126, 277–83.

Michener, C.D. (1974). *The Social Behaviour of Bees*. Cambridge, MA: Belknap Press of Harvard University Press.

Mikhailov, K., Sabath, K. & Kurzanov, S. (1990). Eggs and nests from the Cretaceous of Mongolia. In *Dinosaur Eggs and Babies*, ed. K. Carpenter, K.F. Hirsch & J.R. Horner, pp. 88–115. Cambridge: Cambridge University Press.

Mobley, J.A. & Prum, R.O. (1995). Phylogenetic relationships of the cinnamon tyrant, *Neopipo cinnamomea*, to the tyrant flycatchers (Tyrannidae). *Condor* 97, 650–62.

Møller, A.P. (1982a). Clutch size in relation to nest size in the swallow *Hirundo rustica*. *Ibis* 124, 339–43.

Møller, A.P. (1982b). Ringduens *Columba palumbus* rede: størrelse og sammenaetning. *Dansk Ornithologisk Forenings Tidsskrift* 76, 123–8.

Møller, A.P. (1984). On the use of feathers in birds' nests: predictions and tests. *Ornis Scandinavica* 15, 38–42.

Møller, A.P. (1987a). Egg predation as a selective factor in nest design: an experiment. *Oikos* 50, 91–4.

Møller, A.P. (1987b). Nest lining in relation to nesting cycle in the swallow *Hirundo rustica*. *Ornis Scandinavica* 18, 148–9.

Møller, A.P. (1990a). Nest predation selects for small nest size in the blackbird. *Oikos* 57, 237–40.

Møller, A.P. (1990b). Effects of parasitism by a haematophagous mite on reproduction in the barn swallow. *Ecology* 71, 2345–57.

Møller, A.P. (1991a). The effect of feather nest lining on reproduction in the swallow *Hirundo rustica*. *Ornis Scandinavica* 22, 396–400.

Møller, A.P. (1991b). Ectoparasite loads affect optimal clutch size in swallows. *Functional Ecology* 5, 351–9.

Møller, A.P. (1992). Female swallow preference for symmetrical male sexual ornaments. *Nature* 357, 238–40.

Møller, A.P. (1993). Female preference for apparently symmetrical male sexual ornaments in the barn swallow *Hirundo rustica*. *Behavioural Ecology and Sociobiology* 32, 371–6.

Møller, A.P. (1994). Parasites as an environmental component of reproduction in birds as exemplified by the swallow *Hirundo rustica*. *Ardea* 82, 161–72.

Monaghan, P., Bolton, M. & Houston, D. (1995). Egg production constraints and the evolution of avian clutch size. *Proceedings of the Royal Society of London*, B 259, 189–91.

Monaghan, P. & Náger, R. (1997). Why don't birds lay more eggs? *Trends in Ecology and Evolution* 12, 270–4.

Moratalla, J.J. & Powell, J.E. (1990). Dinosaur nesting patterns. In *Dinosaur Eggs and Babies*, ed. K. Carpenter, K.F. Hirsch & J.R. Horner, pp. 37–46. Cambridge: Cambridge University Press.

Moreau, R.E. (1942). Nesting of African birds in association with other living things. *Ibis* 84, 240–63.

Moreau, R.E. (1944). Clutch size: a comparative study, with special reference to African birds. *Ibis* 86, 286–347.

Moreno, J., Bustamante, J. & Viñuela, J. (1995). Nest maintenance and stone theft in the chinstrap penguin (*Pygoscelis antarctica*): 1. Sex roles and effects on fitness. *Polar Biology* 15, 533–40.

Moreno, J., Soler, M., Møller, A.P. & Linden, M. (1994). The function of stone carrying in the black wheatear, *Oenanthe leucura*. *Animal Behaviour* 47, 1297–309.

Moss, W.W. & Camin, J.H. (1970). Nest parasitism, productivity, and clutch size in purple martins. *Science* 168, 1000–3.

Mourer-Chauviré, C. & Poplin F. (1985). Le mystère des tumulus de

Nouvelle-Calédonie. *La Recherche* 16, 1094.

Myers, J.G. (1935). Nesting associations of birds with social insects. *Transactions of the Royal Entomological Society, London* 83, 11–22.

Nagy, K.A. (1980). CO_2 production in animals: analysis of potential errors in the doubly labeled water method. *American Journal of Physiology* 238, R466–73.

Nagy, K.A., Siegfried, W.R. & Wilson, R.P. (1984). Energy utilisation in free-ranging jackass penguins, *Spheniscus demersus. Ecology* 65, 1648–55.

Nelson, J.B. (1978). *The Sulidae Gannets and Boobies.* Oxford: Oxford University Press.

New Scientist (1988). New high for termites. *New Scientist* 118 (1601), 30.

Nickell, W.P. (1958). Variations in the engineering features of the nests of several species of birds in relation to nest sites and nesting materials. *Butler University Botanical Studies* 13, 121–40.

Nilsson, J-Å. & Svensson, E. (1996). The cost of reproduction: a new link between current reproductive effort and future reproductive success. *Proceedings of the Royal Society London, B* 263, 711–14.

Nilsson, S.G. (1986). Evolution of hole-nesting in birds: on balancing selection pressures. *Auk* 103, 432–5.

Nilsson, S.G., Johnsson, K. & Tjernberg, M. (1991). Is avoidance by black woodpeckers of old nest holes due to predators? *Animal Behaviour* 41, 439–41.

Nisbet, I.C.T. (1977). Courtship feeding and clutch size in common terns *Sterna hirundo.* In *Evolutionary Ecology*, ed. B. Stonehouse & C.M. Perrins, pp. 101–9. London: Macmillan.

Nolte, K.R. & Fulbright, T.E. (1996). Nesting ecology of scissor-tailed flycatchers in South Texas. *Wilson Bulletin* 108, 302–16.

Norell, M.A., Clark, J.M., Chiappe, L.M. & Dashzeveg, D. (1995). A nesting dinosaur. *Nature* 378, 774–6.

Nores, A.I. & Nores, M. (1994a). Nest building and nesting behaviour of the brown cacholote. *Wilson Bulletin* 106, 106–20.

Nores, A.I. & Nores, M. (1994b). Old nest accumulation as anti-predator strategy in brown cacholotes. *Journal für Ornithologie* 135, 198.

North, A.J. (1901–4). *Nests and Eggs of Birds Found Breeding in Australia and Tasmania*, Vol. 1. Sydney: Australian Museum.

Nottebohm, F. (1980). Brain pathways for vocal learning in birds; a review of the first ten years. In *Progress in Psychology and Psychological Psychology*, Vol. 9, ed. J.M. Sprage & A.N. Epstein, pp. 85–124. New York: Academic Press.

Nur, N. (1988). The costs of reproduction in birds: an examination of the evidence. *Ardea* 76, 155–68.

Ogden, C.G. & Hedley, R.H. (1980). *An Atlas of Freshwater Testate Amoebae.* Oxford: British Museum (Natural History), Oxford University Press.

Ogilvie, C.M. (1951). The building of a rookery. *British Birds* 44, 1–5.

O'Hear, A. (1997). *Beyond Evolution*. Oxford: Clarendon Press.

Oliver, W.R.B. (1955). *New Zealand Birds*, 2nd edn. Wellington: A.H. & A.W. Reed.

Olsson, K. & Allander, K. (1995). Do fleas, and/or old nest material, influence nest site preference in hole nesting passerines? *Ethology* 101, 160–70.

Oniki, S. (1979). Is nesting success of birds low in the tropics? *Biotropica* 11, 60–9.

Oniki, Y. (1975). The behaviour and ecology of slaty antshrikes (*Thamnophilus punctatus*) on Barro Colorado Island, Panamá Canal Zone. *Anais Academia Brasileira de Ciências* 47, 477–515.

Oniki, Y. (1985). Why robin eggs are blue and birds build nests: statistical tests for Amazonian birds. In *Neotropical Ornithology. Ornithological Monographs No. 36*, ed. P.A. Buckley, M.S. Foster, E.S. Morton, R.S. Ridgely & F.G. Buckley, pp. 536–45. Washington, DC: American Ornithologists' Union.

Opell, B.D. (1994). The ability of spider cribellar prey capture thread to hold insects with different surface features. *Functional Ecology* 8, 145–50.

Oppliger, A., Richner, H. & Christe, P. (1994). Effect of an ectoparasite on lay date, nest-site choice, desertion, and hatching success in the great tit (*Parus major*). *Behavioural Ecology* 5, 130–4.

Padian, K. (1998). When is a bird not a bird. *Nature* 393, 729–30.

Padian, K. & Chiappe, L.M. (1998). The origin and early evolution of birds. *Biological Reviews* 73, 1–42.

Parker, G.A. (1983). Mate quality and mating decisions. In *Mate Choice*, ed. P. Bateson, pp. 141–66. New York: Cambridge University Press.

Parker, S. & Gibson, K.R. (1979). A developmental model for the evolution of language and intelligence in early hominids. *Behavioural and Brain Sciences* 2, 367–408.

Pateff, P. (1926). Fortflanzungeserscheenungen bei *Difflugia mammillaris* Penard und *Clypeolina marginata* Penard. *Archiv für Protistenkunde* 55, 516–44.

Patterson, I.J. (1965). Timing and spacing of broods in black-headed gulls *Larus ridibundus*. *Ibis* 107, 433–59.

Peakall, D.B. (1960). Nest records of the yellowhammer. *Bird Study* 7, 94–103.

Pearson, T.G. & Burroughs, J. (1936). *Birds of America*. New York: Garden City Publishing Company Inc.

Peckover, W.S. (1970). The fawn-breasted bowerbird (*Chlamydera cerviniventris*). *Proceedings of the 1969 Papua New Guinea Science Society* 21, 23–35.

Perrins, C.M. (1979). *British Tits*. London: Collins, New Naturalist.

Petersen, K.L. & Best, L.B. (1985a). Nest-site selection by sage sparrows. *Condor*, 87, 217–21.

Petersen, K.L. & Best, L.B. (1985b). Brewer's sparrow nest-site characteristic in a sagebrush community. *Journal of Field Ornithology* 56, 23–7.

Petit, L.J. (1991). Adaptive tolerance of cowbird parasitism by prothonotary warblers: a consequence of nest-site limitation? *Animal Behaviour* 41, 425–32.

Petrie, M. (1994). Improved growth and survival of offspring of peacocks with more elaborate trains. *Nature* 371, 598–9.

Pettifor, R.A. (1993a). Brood-manipulation experiments I. The number of offspring surviving per nest in blue tits (*Parus caeruleus*). *Journal of Animal Ecology* 62, 131–44.

Pettifor, R.A. (1993b). Brood-manipulation experiments II. A cost of reproduction in blue tits (*Parus caeruleus*). *Journal of Animal Ecology* 62, 145–59.

Pettifor, R.A., Perrins, C.M. & McCleery, R.H. (1988). Individual optimisation of clutch size in great tits. *Nature* 336, 160–2.

Picman, J. (1988). Experimental study of predation on eggs of the ground-nesting birds: effects of habitat and nest distribution. *Condor* 90, 124–31.

Poulton, E.B. (1931). Association of birds' nests with nests of insects. *Proceedings of the Royal Entomological Society, London* 4, 88–90.

Prestwich, K.N. (1977). The energetics of web building in spiders. *Comparative Biochemistry and Physiology* 57A, 321–6.

Pruett-Jones, M.A. & Pruett-Jones, S.G. (1982). Spacing and distribution of bowers in Macgregor's bowerbird (*Amblyornis macgregoriae*). *Behavioural Ecology and Sociobiology* 11, 25–32.

Pruett-Jones, S.G. & Pruett-Jones, M.A. (1988). The use of court objects by Lawes's parotia. *Condor* 90, 538–45.

Prum, R.O. & Lanyon, W.E. (1989). Morphology and phylogeny of the *Schiffornis* group (Tyrannoidea). *Condor* 91, 444–61.

Quiang, J., Currie, P.J., Norell, M.A. & Shu-An, J. (1998). Two feathered dinosaurs from northeastern China. *Nature* 393, 753–61.

Raven, C.E. (1950). *John Ray Naturalist; His Life and Works*. Cambridge: Cambridge University Press.

Riakow, R.J. (1986). Why are there so many kinds of passerine birds? *Systematic Zoology* 35, 255–9.

Richner, H., Oppliger, A. & Christe, P. (1993). Effect of an ectoparasite on reproduction in great tits. *Journal of Animal Ecology* 62, 703–10.

Ricklefs, R.E. (1969). An analysis of nesting mortality in birds. *Smithsonian Contribution to Zoology* No. 9, 1–48. Washington: Smithsonian Institution Press.

Ricklefs, R.E. & Hainsworth, F.R. (1968). Temperature regulation in nestling cactus wrens: the development of homeothermy. *Condor* 70, 121–7.

Ricklefs, R.E. & Starck, J.M. (1996). Applications of phylogenetically independent contrasts: a mixed progress report. *Oikos* 77, 167–72.

Rivera-Milán, F.F. (1996). Nest density and success of Columbids in Puerto Rico. *Condor* 98, 100–13.

Robinson, M.H. (1981). A stick is a stick and not worth eating: on the definition of mimicry. *Biological Journal of the Linnaean Society* 16, 15–20.

Robinson, M.H. & Robinson, B. (1972). The structure, possible function and origin of the remarkable ladder web built by a New Guinea orb-web spider (Araneae – Arachnidae). *Journal of Natural History* 6, 687–94.

Robinson, S.K. (1986). Competitive and mutualistic interactions among females in a neotropical oriole. *Animal Behaviour* 34, 113–22.

Röell, A. & Bossema, I. (1982). A comparison of nest defence by jackdaws, rooks, magpies and crows. *Behavioural Ecology and Sociobiology* 11, 1–6.

Rohwer, F.C. & Freeman, S. (1989). The distribution of conspecific nest parasitism in birds. *Canadian Journal of Zoology* 67, 239–53.

Roper, J.J. & Goldstein, R.R. (1997). A test of the Skutch hypothesis: does activity at nests increase nest predation risk? *Journal of Animal Biology* 28, 111–16.

Roper, T.J. (1992). Badger *Meles meles* setts: architecture, internal environment and function. *Mammalian Review* 22, 43–53.

Rose, K.D. & Emry, R.J. (1983). Extraordinary fossorial adaptations in the Oligocene *Epoicotherium* and *Xenocranium* (Mammalia). *Journal of Morphology* 175, 33–56.

Rosenthal, G.A. & Janzen, D.J. (eds.) (1979). *Herbivores. Their Interaction with Secondary Metabolites.* New York: Academic Press.

Rosenzweig, M.L. (1996). Colonial birds probably do speciate faster. *Evolutionary Ecology* 10, 681–3.

Rowley, I. (1970). The use of mud in nest-building – a review of the incidence and taxonomic importance. *Ostrich*, Suppl. 8, 139–48.

Rowley, I. (1978). Communal activities among white-winged choughs *Corcorax melanorhamphus. Ibis* 120, 178–97.

Rowley, J.S. (1966). Breeding records of birds of the Sierra Madre Del Sur, Oaxaca, Mexico. *Proceedings of the Western Foundation of Vertebrate Zoology* 1 (3), 107–204.

Rutnagur, R. (1990). Nest structure and related building behaviour in the rook *Corvus frugilegus.* Unpublished PhD Thesis, Glasgow University.

Ryan, M.J. (1997). Sexual selection and mate choice. In *Behavioural Ecology: An Evolutionary Approach*, 4th edn, ed. J.R. Krebs & N.B. Davies, pp. 179–202. Oxford: Blackwell Science.

Saether, B-E. (1985). Variation in reproductive traits in European passerines in relation to nest site: allometric scaling to body weight or adaptive variation? *Oecologia* 68, 7–9.

Sakagami, S.F. & Yoshikawa, K. (1968). A new ethospecies of *Stenogaster* wasp from Sarawak with a comment on the value of ethological characters in animal taxonomy. *Annotnes Zoologia Japonica* 41, 77–84.

Samuel, D.E. (1969). House sparrow occupancy of cliff swallow nests. *Wilson Bulletin* 81, 103–04.

Samuel, D.E. (1971). The breeding biology of barn and cliff swallows in West Virginia. *Wilson Bulletin* 83, 284–301.

Sargent, T.D. (1965). The role of experience in the nest building of the zebra finch. *Auk* 82, 48–61.

Sattaur, O. (1991). Termites change the face of Africa. *New Scientist* 129 (1753), 27.

Schaefer, V.H. (1976). Geographic variation in the placement and structure of oriole nests. *Condor* 78, 443–8.

Schmidt, R.S. (1955). Termite (*Apicotermes*) nests – important ethological material. *Behaviour* 8, 344–56.

Scott, V.E., Evans, K.E., Patton, D.R. & Stone, C.P. (1977). *Cavity-nesting Birds of the North American Forests. Agriculture Handbook 511*. Washington, DC: Forest Service, US Department of Agriculture.

Seehausen, O., van Alphen, J.J.M. & Witte, F. (1997). Cichlid fish diversity threatened by eutrophication that curbs sexual selection. *Science* 277, 1808–11.

Sengupta, S. (1981). Adaptive significance of the use of margosa leaves in nests of house sparrows *Passer domesticus. Emu* 81, 114–15.

Seymour, R.S., Withers, D.C. & Weathers, W.W. (1998). Energetics of burrowing, running, and free-living in the *Namib* Desert golden mole (*Eremitalpa namibensis*). *Journal of Zoology, London* 244, 107–17.

Shelly, L.O. (1935). Flickers attacked by starlings. *Auk* 52, 93.

Shields, W.M. & Crook, J.R. (1987). Barn swallow coloniality: a net cost for group breeding in the Adirondacks? *Ecology* 68, 1373–86.

Short, L.L. (1979). Burdens of the picid hole-nesting habit. *Wilson Bulletin* 91, 16–28.

Sibley, C.G., Ahlquist, J.E. & Monroe, B.L. (1988). A classification of the living birds of the world based on DNA–DNA hybridisation studies. *Auk* 105, 409–23.

Sibley, S.G. & Monroe, B.L. (1990). *Distribution and Taxonomy of Birds of the World*. Newhaven: Yale University Press.

Sick, H. (1979). Zür Nistweise der Cotingiden and Zipholena. *Journal für Ornithologie.* 120, 73–7.

Sick, H. (1993). *Birds in Brazil* (Translated from Portuguese by W. Belton). Princeton, NJ: Princeton University Press.

Siegfried, W.R. (1975). On the nest of the hamerkop. *Ostrich* 46, 267.

Simon, E. (1891). Observations biologiques sur les Arachnides. *Annales de la Société Entomologique de France* 60, 5–14.

Simpson, K. & Day, N. (1989). *Field Guide to the Birds of Australia*. London: Christopher Helm.

Skeel, M.A. (1983). Nesting success, density, philopatry, and nest-site selection of the Whimbrel (*Numenius phaeopus*) in different habitats.

Canadian Journal of Zoology 61, 218–25.

Skowron, C. & Kern, M. (1980). The insulation in nests of selected North American songbirds. *Auk* 97, 816–24.

Skutch, A.F. (1954). *Life Histories of Central American Birds: Families Fringillidae, Thraupidae, Icteridae, Parulidae and Coerebidae*, No. 31. Berkeley, CA: Cooper Ornithological Society.

Skutch, A.F. (1960). *Life Histories of Central American Birds*, Vol. 2: *Families Fringillidae, Vireonidae, Sylviidae, Turdidae, Trogloditidae, Paridae, Corvidae, Hirundinidae and Tyrannidae*, No. 34. Berkeley, CA: Cooper Ornithological Society.

Skutch, A.F. (1961). The nest as a dormitory. *Ibis* 103a, 50–70.

Skutch, A.F. (1964). Life history of the blue-diademed motmot *Momotus momota*. *Ibis* 106, 321–32.

Skutch, A.F. (1969). *Life Histories of Central American Birds*, Vol. 3: *Families Cotingidae, Pipridae, Formicariidae, Furnariidae, Dendro-colaptidae and Picidae*, No. 35. Berkeley, CA: Cooper Ornithological Society.

Skutch, A.F. (1973). *The Life of the Hummingbird*. New York: Crown Publishers, Inc.

Skutch, A.F. (1976). *Parent Birds and Their Young*. Austin: University of Texas Press.

Skutch, A.F. (1983). *Birds of Tropical America*. Austin: University of Texas Press.

Skutch, A.F. (1985). Clutch size, nesting success and predation on nests of neotropical birds, reviewed. In *Neotropical Ornithology. Ornithological Monographs* No. 36, ed. P.A. Buckley, M.S. Foster, E.S. Morton, R.S. Ridgely & F.G. Buckley, pp. 575–94. Washington, DC: American Ornithologists' Union.

Slagsvold, T. (1982a). Clutch size variation in passerine birds: the nest predation hypothesis. *Oecologia* 54, 159–69.

Slagsvold, T. (1982b). Clutch size and hatching asynchrony in birds: experiments with the fieldfare (*Turdus pilaris*). *Ecology* 63, 1389–99.

Slagsvold, T. (1984). Clutch size variation of birds in relation to nest predation: on the cost of reproduction. *Journal of Animal Ecology* 53, 945–53.

Slagsvold, T. (1989a). Experiments on clutch size and nest size in passerine birds. *Oecologia* 80, 297–302.

Slagsvold, T. (1989b). On the evolution of clutch size and nest size in passerine birds. *Oecologia* 79, 300–5.

Smith, L.H. (1968a). *The Lyrebird*. Melbourne: Lansdown Press.

Smith, N.G. (1968b). The advantage of being parasitized. *Nature (London)* 219, 690–4.

Smith, N.G. (1980). Some evolutionary, ecological and behavioural correlates of communal nesting by birds with wasps and bees. *Proceedings of the 17th International Congress of Ornithology* 2, 1199–205.

Smith, W.K., Roberts, S.W. & Miller, P.C. (1974). Calculating the nocturnal energy expenditure of an incubating Anna's hummingbird. *Condor* 76, 176–83.

Smythies, B.E. (1986). *The Birds of Burma*, 3rd edn. Liss: Nimrod Press; Ontario: Silvio Mattacchione & Co.

Snow, D.W. (1962). A field study of the black and white manakin, *Manacus manacus* in Trinidad. *Zoologica, New York Zoological Society* 47, 63–103.

Snow, D.W. (1976). *The Web of Adaptation: Bird Studies of the American Tropics*. London: Collins.

Snow, D.W. (1978). The nest as a factor determining clutch-size in tropical birds. *Journal für Ornithologie* 119, 227–230.

Snow, D.W. & Snow, B.K. (1963). Breeding and the annual cycle in three Trinidad thrushes. *Wilson Bulletin* 75, 27–41.

Soler, J.J., Møller, A.P. & Soler, M. (1998). Nest building, sexual selection and parental investment. *Evolutionary Ecology* 12, 427–41.

Soler, J.J., Soler, M., Møller, A.P. & Martinez, J.G. (1995). Does the great spotted cuckoo choose magpie hosts according to their parenting ability? *Behavioural Ecology and Sociobiology* 36, 201–6.

Starck, J.M. (1993). Evolution of avian ontogenies. In *Current Ornithology*, Vol. 10, ed. D.M. Power, pp. 275–366. New York: Plenum Press.

Starck, J.M. & Ricklefs, R.E. (1994). Patterns of development in birds: the altricial–precocial system. *Journal für Ornithologie* 135, 326.

Starck, J.M. & Sutter, E. (1994). Comparative analysis of growth and the evolution of superprecociality in megapodes. *Megapode Newsletter* 8, 11–14.

Starrett, A. (1993). Adaptive resemblance: unifying concept for mimicry and crypsis. *Biological Journal of the Linnaean Society* 48, 299–317.

Stehr, F.W. & Cook, E.F. (1968). A revision of the genus *Malacosoma* Hubner in North America (Lepidoptera: Lasiocampidae): systematics, biology, immatures and parasites. *US National Museum Bulletin* 276, 1–34.

Steyn, P. (1992). Gabar Goshawks and colonial spiders. *Gabar*, 7, 21.

Stone, T. (1989). Origins and environmental significance of shell and earth mounds in Northern Australia. *Archaeology in Oceania* 24, 59–64.

Storer, N.P. & Hansell, M.H. (1992). Specialisation in the choice and use of silk in the nest of the chaffinch (*Fringilla coelebs*) Aves Fringillidae. *Journal of Natural History* 26, 1421–30.

Studer, A. (1994). Analysis of nest success in Brazilian birds. *Journal für Ornithologie* 135(3), 298.

Studer, A. & Vielliard, J. (1990). The nest of the wing-banded hornero *Furnarius figulus* in Northeastern Brazil. *Ararajuba* 1, 39–41.

Thaler, E. (1976). Nest und Nestbau von Winter- und Sommergoldhähnchen (*Regulus regulus und R. ignicappillus*). *Journal für Ornithologie* 117, 121–44.

Thiollay, J.M. (1991). Foraging, home range use and social behaviour of a group-living rainforest raptor, the red-throated caracara *Daptrius americanus*. *Ibis* 133, 382–93.

Tidemann, S.C. & Marples, T.G. (1988). Selection of nest sites by three species of fairy-wrens (*Malurus*). *Emu* 88, 9–15.

Tinbergen, N. (1953). Specialists in nest-building. *Country Life*, January 30, 270–1.

Tomialojc, L. (1992). Colonisation of dry habitats by the song thrush *Turdus philomelos*: is the type of nest material an important nest constraint? *Bulletin of the British Ornithologists' Club* 112, 27–34.

Traylor, M.A. & Fitzpatrick, J.W. (1982). A survey of the tyrant fly-catchers. *Living Bird* 19, 7–50.

Tremblay, J-P., Gautier, G., Lepage, D. & Desrochers, A. (1997). Factors affecting nesting success in greater snow geese: effects of habitat and association with snowy owls. *Wilson Bulletin* 109, 449–61.

Ueta, M. (1994). Azure-winged magpies, *Cyanopica cyana*, 'parasitize' nest defence provided by Japanese lesser sparrowhawks, *Accipiter gularis*. *Animal Behaviour* 48, 871–4.

Ueta, M. (1998). Azure-winged magpies avoid nest predation by nesting near a Japanese lesser sparrowhawk's nest. *Condor* 100, 400–2.

Vander Werf, E. (1992). Lack's clutch size hypothesis: an examination of the evidence using meta-analysis. *Ecology* 73, 1699–1705.

van Lawick-Goodall, J. & van Lawick, H. (1966). Use of tools by the Egyptian vulture *Neophron percnopterus*. *Nature* 212, 1468–9.

Van Someren, V.G.L. (1956). Days with birds. Fieldiana in *Zoology* Vol. 38. Chicago: Chicago Natural History Museum.

Vaurie, C. (1980). Taxonomy and geographical distribution of the Fur-nariidae (Aves, Passeriformes). *Bulletin of the American Museum of Natural History* 166, 1–357.

Vellenga, R.E. (1970). Behaviour of the male satin bower-bird at the bower. *The Australian Birdbander* March 1970, 3–11.

Verner, J. (1964). Evolution of polygamy in the long-billed marsh wren. *Evolution, NY* 18, 252–61.

Verner, J. & Engelsen, G.H. (1970). Territories, multiple nesting, and polygyny in the long-billed marsh wren. *Auk* 87, 557–67.

Vleck, D., Vleck, C.M. & Seymour, R.S. (1984). Energetics of embryonic development in the megapode birds, mallee fowl *Leipola ocellata* and brush turkey *Alectura lathami*. *Physiological Zoology* 57, 444–56.

Vollrath, F. (1992). Spiders webs and silks. *Scientific American* 266 (3), 70–6.

Waldbauer, G.P., Scarbrough, A.G. & Sternburgh, J.G. (1982). The allocation of silk in compact and baggy cocoons of *Hyalophora cecropia*. *Entomologia Experimentalis et Applicata* 31, 191–6.

Wallace, A.R. (1867). The philosophy of birds' nests. *Intellectual Observer* 11, 413–20.

Wallace, A.R. (1889). *Darwinism* (3rd edn 1901). London: Macmillan.

Wallace, J.B. & Sherberger, F.F. (1975). The larval dwelling and feeding structure of *Macronema transversum* (Walker) (Trichoptera: Hydropsychidae). *Animal Behaviour* 23, 592–6.

Walsberg, G.E. (1981). Nest-site selection and the radiative environment of the warbling vireo. *Condor* 83, 86–8.

Walsberg, G.E. & King, J.R. (1978a). The energetic consequences of incubation for two passerine species. *Auk* 95, 644–55.

Walsbeg, G.E. & King, J.R. (1978b). The heat budget of incubating mountain white-crowned sparrows (*Zonotrichia leucophrys oriantha*) in Oregon. *Physiological Zoology* 51, 92–103.

Warham, J. (1957). Notes on the display and behaviour of the great bower-bird. *Emu* 57, 73–8.

Washburn, S.C. (1959). Speculations on the interrelations of the history of tools and biological evolution. *Human Biology* 31, 21–31.

Wasylik, A. (1971). Nest types and the abundance of mites. *Ekologia Polska* 19, 689–99.

Watson, J.P. (1967). A termite mound in an iron age burial ground in Rhodesia. *Journal of Ecology* 55, 663–9.

Weaver, C.M. (1982). Breeding habitats and status of the golden-shouldered parrot *Psephotus chrysopterygius*, in Queensland. *Emu* 82, 2–6.

Weeks, H.P. (1977). Nest reciprocity in eastern phoebes and barn swallows. *Wilson Bulletin* 89, 632–5.

Weeks, H.P. (1978). Clutch size variation in the eastern phoebe in Southern Indiana. *Auk* 95, 656–66.

Wenzel, J. (1991). Evolution of nest architecture. In *The Social Biology of Wasps*, ed. K.G. Ross & R.W. Matthews, pp. 480–519. Ithaca: Cornell University Press.

Wenzel, J. (1993). Application of the biogenetic law to behavioural ontogeny: a test using nest architecture in paper wasps. *Journal of Evolutionary Biology* 6, 229–47.

Wenzel, J. (1996). Learning, behaviour programmes and higher level values in the nest construction of *Polistes*. In *Natural History and Evolution of Paper Wasps*, ed. S. Turillazzi & M.J. West-Eberhard, pp. 58–74. Oxford: Oxford University Press.

Wesolowski, T. (1994). On the origins of parental care and early evolution of male and female parental roles in birds. *American Naturalist* 143, 39–58.

Wetmore, A. (1972). The birds of the Republic of Panama, Part 3. *Smithsonian Miscellaneous Collections* 150, 1–631.

Wheelwright, N.T., Lawler, J.J. & Weinstein, J.H. (1997). Nest-site selection in savannah sparrows: using gulls as scarecrows? *Animal Behaviour* 53, 197–208.

Whiten, A. & Byrne, R.W. (1988). The Machiavellian intelligence hypotheses: editorial. In *Machiavellian Intelligence. Social Expertise and the Evolution of Intellect in Monkeys, Apes and Humans*, ed. R.W. Byrne & A. Whiten, pp. 1–9. Oxford: Clarendon Press.

Willson, M.F. (1974). Avian community organisation and habitat structure. *Ecology* 55, 1017–29.

Wilson, E.O. (1971). *The Insect Societies*. Cambridge, MA: Harvard University Press, Belknap.

Wimberger, P.H. (1984). The use of green plant material in bird nests to avoid ectoparasites. *Auk* 101, 615–18.

Winkler, D.W. (1993). Use and importance of feathers as nest lining in tree swallows (*Tachycineta bicolor*). *Auk* 110, 29–36.

Winkler, D.W. & Sheldon, F.H. (1993). Evolution of nest construction in swallows (Hirundinidae): a molecular phylogenetic perspective. *Proceedings of the National Academy of Sciences USA* 90, 5705–7.

With, K.A. (1994). The hazards of nesting near shrubs for a grassland bird, the McCown's longspur. *Condor* 96, 1009–19.

Withers, P.C. (1977). Energetic aspects of reproduction by the cliff swallow. *Auk* 94, 718–25.

Wittenberger, J.F. & Hunt, G.L. (1985). The adaptive significance of coloniality in birds. In *Avian Biology*, Vol. VIII, ed. D.S. Farner, J.S. King & K.C. Parkes, pp. 1–58. London: Academic Press.

Wunderle, J.M. & Pollock, K.H. (1985). The bananaquit-wasp nesting association and a random choice model. In *Neotropical Ornithology*. *Ornithological Monographs* No. 36, ed. P.A. Buckley, M.S. Foster, E.S. Morton, R.S. Ridgely & F.G. Buckley, pp. 595–603. Washington, DC: American Ornithologists' Union.

Yamauchi, A. (1993). Theory of intraspecific nest parasitism in birds. *Animal Behaviour* 46, 335–45.

Yanes, M., Herranz, J. & Suarez, F. (1996). Facultative nest-parasitism among Iberian shrubsteppe passerines. *Bird Study* 43, 119–23.

Yom-Tov, Y. (1980). Intraspecific nest parasitism in birds. *Biological Reviews* 55, 93–108.

Yosef, R. (1992). From nest building to fledging of young in great grey shrikes (*Lanius excubitor*) at Sede Boqer, Israel. *Journal für Ornithologie* 133, 279–85.

Young, B.E., Kaspari, M. & Martin, T.E. (1990). Species-specific nest site selection by birds in ant-acacia trees. *Biotropica* 22, 310–15.

Zahavi, A. (1975). Mate selection – a selection for a handicap. *Journal of Theoretical Biology* 53, 205–14.

Zann, R. & Rossetto, M. (1991). Zebra finch incubation: brood patch, egg temperature and thermal properties of the nest. *Emu* 91, 107–20.

Zyskowski, K. & Prum, R.O. (1999). Phylogenetic analysis of the nest architecture of neotropical ovenbirds (Furnariidae). *Auk* 116, 891–911.

Author index

General index

Species index